Praise for
Mystery Teachings from the Living Earth

"*Mystery Teachings from the Living Earth* is the clearest, simplest, and most relevant description of the ancient Mystery wisdom that I have read in many years. It is filled with rich and important insights that amply confirm Greer's place as one of the premier spiritual teachers of our time. It is a gift of illumination for every serious spiritual seeker."

—David Spangler, author of *Apprenticed to Spirit,*
The Laws of Manifestation, and
Subtle Worlds: An Explorer's Field Guide

"This is an important book. A really important book. Everyone who wants to know about the relationship between the Mysteries and the earth on which we live should read it. Those who believe in the importance of ecology—especially spiritually-based ecology—must read it. A timely and valuable book."

—John Matthews, co-author of *The Wildwood Tarot*

"A masterly analysis and description of what genuine mystery schools and magical practitioners are all about, by one who is obviously as wise as he is well experienced, presenting ancient teachings in a form equal to the challenges of the world today. Very practical, too. Highly recommended!"

—Gareth Knight, author of
A Practical Guide to Qabalistic Symbolism

"A wise, graceful, and eloquent introduction to some of reality's deepest truths. It deserves to become a bestseller."

—Richard Smoley, author of *Inner Christianity* and *Forbidden Faith: The Secret History of Gnosticism*

"*Mystery Teachings from the Living Earth* alchemically renders a third element beyond New Age platitudes such as 'we create our own reality,' and nihilistic notions of meaningless absurdity that overwhelm the soul and mire us in despair. Greer masterfully demonstrates that the planetary initiation in which we are deeply embroiled compels us to embrace an ecology of spirit and evolve from egocentric self-destruction toward ecocentric wisdom and wholeness."

—Carolyn Baker, PhD, author of *Navigating the Coming Chaos: A Handbook for Inner Transition* and *Sacred Demise: Walking the Spiritual Path of Industrial Civilization's Collapse*

"What an elegant work! Greer moves gracefully between the keenly observed natural world and the deepest esoteric mysteries to present *useable* teachings—bright, vibrant, and wonderfully pragmatic—as satisfying for material-plane use as they are for spirit. His recurring motif of the meadow is brilliant, as the tiniest workings of animal, vegetable, and mineral life expand metaphorically to bring these vast metaphysical themes to life. This is just what I've been hungering for!"

—Renna Shesso, author of *A Magical Tour of the Night Sky*

"John Michael Greer, whose many works have demonstrated his well-developed scholarly aptitude, articulates here a pragmatic, mystical, and conscientious set of principles as a practitioner-leader of a contemporary mystery school. His spiritual ecology, critical analysis of cultural excess, ethical guidelines, and corrective reconstruction of 'ancient teachings' is inspirational, balanced, and accessible. The emphasis falls on self-discipline, clear thinking, and long-term commitment to spiritual goals for the improvement of human life; this is not a book about glamour or power, but intelligence, refinement, and magical self-discovery. If you want a foundational introduction to the mysteries, read this book!"

—Lee Irwin, Professor of Religious Studies,
College of Charleston

MYSTERY TEACHINGS FROM THE LIVING EARTH

MYSTERY
TEACHINGS
from the
LIVING EARTH

an introduction to spiritual ecology

JOHN MICHAEL GREER

WEISER BOOKS
San Francisco, CA / Newburyport, MA

First published in 2012 by Weiser Books, an imprint of

Red Wheel/Weiser, LLC
With offices at:
665 Third Street, Suite 400
San Francisco, CA 94107
www.redwheelweiser.com

ISBN: 978-1-57863-489-7

Library of Congress Cataloging-in-Publication Data is available upon request.

Cover design by Jim Warner
Cover photograph: Serpent Mound State Memorial,
Locust Grove, Ohio © *Tony Linck /SuperStock*
Interior by Kathryn Sky Peck
Typeset in Sabon

Printed in the United States of America
MAL
10 9 8 7 6 5 4 3 2 1

The paper used in this publication meets the minimum requirements of the
American National Standard for Information Sciences—Permanence of Paper
for Printed Library Materials Z39.48-1992 (R1997).

CONTENTS

INTRODUCTION

Over the last thirty years or so, a particular set of teachings about human potential and the nature of the universe has become wildly popular across much of the world. These teachings have often been presented as the authentic secrets of ancient, secret mystery schools, but it's hard to think of anything less secret and more up to date. If you've read books, watched videos, visited websites, attended workshops, or simply tuned into the zeitgeist by way of casual conversations, you're almost certain to have encountered them.

These teachings make some remarkable claims. They insist that each of us creates his or her own reality by means of our attitudes and thoughts, and so we can have whatever reality we want if we simply exchange limiting attitudes and negative thoughts for liberating attitudes and positive thoughts. They argue that the only limits the universe contains are the ones imposed on it by the human mind, and that everything—including material wealth—exists in infinite amounts, just waiting for each of us to claim as big a share as we wish. They claim that the universe is governed by a "law of attraction" that allows each of us to attract whatever we want from the world around us, in the confident expectation that it will show up promptly on cue. They announce, finally, that those who embrace these ideas and undergo exotic initiations based on them can pass through a wondrous process of ascension that will liberate them from death and every other limitation, and that the entire world is on the brink of a cataclysmic change that will result in a Utopian future free of everything about the world that most of us dislike.

These teachings paint a very flattering picture of humanity's nature and powers, and it is no wonder that they have attracted plenty of attention. They also have a certain basis in human experience, since attitudes and expectations do have a great deal of influence over our experience of the world. Still, there are at least two problems with these appealing ideas. The first is that despite the claims made for them, they are not the teachings of the mystery schools. (This can be looked up easily enough, since the mystery schools still exist and have put their teachings into print many times over the last few centuries.) Still, the second and more immediate problem is that the teachings simply don't work as advertised.

The most recent round of economic troubles has provided as much evidence for this last statement as anyone could wish. Hundreds of thousands, perhaps millions, of people set out to use the "law of attraction" and similar methods of creating their own reality through positive thinking to become permanently wealthy by buying and selling real estate. If you know any of them, you know how well that worked. Some of the people who marketed the teachings through books, videos, seminars, and the like did very well indeed, to be sure, but nearly all those who set out to put the teachings into practice—whether they tried to get the universe to give them a seven-figure income, or paid for workshop after workshop in the hope of becoming an ascended master, or waited eagerly for the world of hard limits and hard knocks to dissolve in apocalyptic fireworks that would usher in the bright new world they had been promised—came out the other end of the process with less money and more troubles than they had when they started.

There are good reasons for this awkward turn of events. In fact, there is a rich irony in the fact that these teachings surfaced when they did. We live in an age in which it is becoming ever harder to ignore the terrible consequences of treating the universe as though it exists solely for human benefit. The rising spiral of environmental crises that fills so many headlines today is being driven by precisely the same attitudes that these teachings present as spiritual wisdom. What causes oil spills and climate change, the devastation of the rain forests and the extinction of species, but the same belief that the universe is somehow obligated to give us whatever we want to take from it?

The authentic teachings of the mystery schools offer a profoundly different way of making sense of the universe and our place in it.

The authentic teachings of the mystery schools offer a profoundly different way of making sense of the universe and our place in it, rooted in a deep knowledge of whole systems—a way that can, without exaggeration, be described as "spiritual ecology." This way of knowledge makes it possible to understand both the real possibilities and the equally real limits that define human existence. At the present time, as one age of the world gives way slowly to another, these ancient teachings are destined to find new expressions that are suited to the possibilities and challenges brought by the new turn of time's great wheel.

This book is an attempt at a new expression of this kind. It makes no claim to authority. It simply offers one way of understanding certain basic truths about existence that have been the heritage of every mystery school since before recorded history began.

John Michael Greer
Cumberland, Maryland
October 15, 2010

AN ECOLOGY OF SPIRIT

Since the most ancient times on record, in every part of the world, a body of traditional wisdom about the nature of the universe and of humanity has been handed down from teacher to student. This wisdom has taken diverse forms and has borrowed the language and symbolism of many creeds and cultures. In all its many forms, however, it deals with the deepest mysteries of human existence: the nature, origin, and destiny of the universe and of the human soul; the hidden powers that lie concealed in humanity; and the practices by which these powers can be awakened, trained, and put to constructive uses. This ancient wisdom has carried many names on its journey down through the centuries; nowadays, in the modern industrial West, it is most commonly known as the mystery teachings.

There is every reason to think that this ancient wisdom had already reached its full richness and complexity long before the first cities emerged from tribal villages and the habit of painting images on wood and cloth gave rise to the first systems of writing. Certainly the oldest forms of the mystery teachings that have come down to us are every bit as rich and subtle as those in circulation today. Where and when the mystery teachings began is a question no one can answer with any assurance. Legends from various corners of the world make many different claims, none of which can be proven or disproven. Still, we do know that the wise men and women who passed down the oldest known mystery teachings were careful students of nature, and it may be that the forgotten founders of the mystery traditions drew their knowledge from the same source: learning the lessons that nature has to teach and phrasing them in whatever ways their students would be most likely to learn and remember.

From ancient times straight through to the present, these mystery teachings were the public face of the mystery schools themselves. A great deal of confusion and misinformation has gathered around these schools in recent years. Some stories about them amount to little more than gaudy fantasies about immortal masters with superhuman powers hidden away in Himalayan fastnesses; others are equally gaudy horror tales about devil worship or political conspiracy. The truth is far simpler and far more interesting.

The mystery schools are, first and foremost, schools. They have their teachers, who have qualified for their positions through many years of studying and practicing the mystery teachings. They have their students, who have qualified to enter the schools by passing certain tests of character and willingness to learn, and who study and practice the teachings of the school. They have their curriculums, which include the study of symbolism, philosophy, and ethics, as well as the practice of inner disciplines such as meditation, ritual, and prayer. The studies are meant to train the students to think clearly, to act wisely, and to make sense of the experiences gained through the practices; the practices themselves are designed to open up states of consciousness unfamiliar to most people outside the mystery traditions, for it is in these states of consciousness that the hidden powers of the human soul are awakened.

The mystery schools also confer ceremonies of initiation. A great deal of nonsense has been spread about these ceremonies over the years by people who have never experienced them. In reality, initiation ceremonies are simply dramatic performances that present core insights of the mystery teachings in symbolic form. In an initiation ceremony, the candidate is shown the symbols and presented with the teachings that are central to the grade of learning he or she is about to enter, and he or she makes a formal promise to accept the duties and pursue the studies of that grade. The course of study and practice that follows the initiation, not the mere fact of passing through such a ceremony, confers whatever inner capacities the grade of initiation is meant to develop in the student.

Even in ancient times, there were already several different mystery schools in most nations, and some of the most famous initiates of those days won much of their fame by traveling from one center of initiation to another, passing through the courses of training offered by different mystery schools and mastering the studies and practices of each. Nowadays travel across long distances is easier and quicker than it was in the days when it might take a month of travel on foot, and another month or more on a sailing vessel, to go from the mystery temples of

Egypt to those of Greece. This has created an even greater diversity of mystery schools in most modern countries. There is no central authority to which the various mystery schools answer; they function independently of one another and present their own teachings to their own students in their own ways.

All this, at least in modern times, is done very quietly and with a minimum of publicity. The work of the mystery schools is carried on in small groups that meet in private homes or rented offices, or by correspondence lessons circulated by mail or the Internet. The schools rarely publicize themselves or their activities, though a determined student who seeks training in the mysteries rarely has to search long before he or she finds a source of instruction.

Now and again it becomes possible for a mystery school to establish a center where the lessons can be taught and the rituals performed on a grander scale, but this has always had certain risks. Human ignorance, which always fears what it does not understand and always tries to destroy what it fears, is a danger that the mysteries and their initiates have always faced; there is also the danger that a mystery school that becomes too prominent will attract people more interested in status and wealth than in the teachings themselves. The cost of maintaining a large complex of buildings, staff, and the like is also not a small factor; in order to make their teachings available to as many people as possible, mystery schools generally ask for money only to the extent that is necessary to cover expenses and, thus, only in very modest amounts.

Any organization that demands substantial sums for teaching and initiation is engaged in the business of making money, not in the work of the mysteries.

Any organization that demands substantial sums for teaching and initiation is engaged in the business of making money, not in the work of the mysteries.

While the schools are usually private, or at most semipublic, their introductory teachings have generally been made public whenever this has been possible. One sign that mystery schools are active and accepting students in any society is the appearance of the basic mystery teachings in some publicly accessible form. These public teachings may use the language of poetry, philosophy, or religion; they often borrow the commonplaces of popular culture in the place and time where they appear, but they always express a distinctive set of attitudes toward humanity and the universe; they always hint at the existence of unrecognized powers lying dormant in the human soul, which can be awakened by paths of discipline and devotion; and they always provide some basic instruction in awakening those powers and pursuing the path of the mysteries.

These public presentations of the basic teachings of the mysteries have a dual purpose. Their obvious purpose is to attract the attention of those who might be suited to the study of the mysteries, to give them some basic instructions, and to encourage them to seek the schools where further instruction can be obtained. Their less obvious purpose is to place certain ideas in circulation in society at large, to help remedy whatever imbalances have crept into the popular thinking of the time, and to further the cause of public education and enlightenment.

In the Western world, during recent times, it has become a common practice to present the core teachings of the mysteries in the form of seven clearly defined laws or principles and then to expand on these principles by way of commentary. While the number of principles is fixed at seven for symbolic reasons, the principles themselves vary from one presentation to another. Every expression of the mystery teachings is shaped by the needs of the time in which it finds its way into the public sphere, and it must rely on the resources of language and knowledge available to its authors and its audience.

These resources change with the passing years. Words lose and gain meanings over time, and ways of using language that communicate clearly to one generation often seem difficult and obscure to the generation that follows. Nor will images drawn from the everyday life of one generation necessarily make sense to those who live in a different age.

In ancient Greece, for example, chariots were an everyday part of life. Most people had either driven one or ridden in one, and everyone knew what happened when the two horses who pulled a chariot decided to go in two different directions at the same time. Plato, the Greek philosopher whose writings presented the mystery teachings to the general reading public of his country and time, could compare the soul to a chariot pulled by unruly horses and get instant understanding from his students and readers. That understanding comes harder to people of our time, most of whom have never seen a chariot outside books or movies.

Changes in knowledge and culture also play their part. In ancient Greece, geometry was a new and exciting science, still being developed by some of the greatest minds of the age, and most people in the Greek city-states had heard at least a little about it. Plato deliberately made use of this when he wrote about the mystery teachings, weaving geometrical patterns into his dialogues and borrowing geometrical metaphors to make his points more clear to his students and readers. At the time, it made his writings accessible to his readers; a few centuries later, another author, Theon of Smyrna, had to write a book titled *The Mathematics Necessary to Understand Plato* because times had changed and ideas that were part of the common currency of thought in Plato's time were no longer familiar.

Every expression of the mystery teachings is shaped by the needs of the time in which it finds its way into the public sphere.

In the early twentieth century, when most versions of the mystery teachings that are in common circulation nowadays were written, psychology was the new and exciting science about which most people had heard a little, and so the authors of these later presentations organized their works according to psychological principles and borrowed psychological examples to pass on the same teachings as Plato's works, in new and more easily accessible forms, to their students and readers. With the passing of time, in turn, many of the psychological references in these writings have become much less familiar than they were, though no Theon of Smyrna has yet emerged to write *The Psychology Necessary to Understand The Kybalion*.

Today, however, ecology—the science of whole systems in nature—occupies the same cutting-edge position that psychology had a century ago and geometry had in Plato's time, and most people know at least a little about it. Ecology is also the most necessary of all sciences today. The survival of our civilization, and possibly of our species as well, depends on how well we can learn to live in harmony with nature at this point in history, and the science of ecology offers crucial guidance in that quest.

As a way of talking about the mystery teachings, furthermore, ecology has a special advantage, because the mystery teachings themselves are a science of whole systems. It is not going too far to call the mystery teachings an ecology of the spirit, just as the science of ecology could well be called the mystery teachings of nature. One consequence of this pattern of parallel meanings is that the teachings of the mysteries can be clarified and mistakes in understanding them put right by comparing them to the ways that whole systems function in the world of nature, where the relationships between whole systems and their parts can be clearly seen and measured.

This comparison allows a clarity that the last generation of presentations of the mystery teachings did not always allow. Those presentations, as already mentioned, drew their structure and symbolism from the science of psychology. That was a wise decision at the time, for popular ideas at the beginning of the twentieth century had come to treat the human mind as little more than a machine, and the mystery schools of that time found it necessary to give their students a higher and more accurate sense of the mind's potentials. Over the years that followed, however, misunderstandings crept in, subtle distinctions were lost, and ideas were embraced in a one-sided way; the important teaching that the mind participates in the creation of the reality it experiences, for example, was thus distorted into the half-truth that "you create your own reality."

It is not going too far to call the mystery teachings an ecology of the spirit, just as the science of ecology could well be called the mystery teachings of nature.

This process of distortion is a familiar reality for students of the mysteries. It happens every century or two and gives rise to confusions of a sort that can be found all through popular spirituality today. People are always eager for teachings that tell them what they already want to hear, and in an age obsessed by the craving for material wealth, it was inevitable that those scraps of the mystery teachings that seem to promise the fulfillment of every material want would be taken out of context, reshaped to fit the ordinary habits of the unawakened mind, and pasted together into a set of ideas that essentially treat the cosmos as though it was some sort of infinite Internet store that never gets around to sending the bill.

The gap between the resulting teachings and those of the authentic mysteries can be measured by the roles of greed and fear, the two great rulers of the unawakened consciousness. Those teachings that fixate on finding ways to get material wealth without earning it show the presence of greed; those teachings that insist that the normal and inevitable human experiences of suffering and death are unreal or unnatural betray the presence of fear. It is useful to compare both these traits with authentic mystery teachings, which recognize that the pursuit of material wealth becomes an obstacle to spirituality when taken beyond the point of meeting one's basic needs and serving the common good, and which also recognize that suffering and death are among the greatest initiations of human existence and are to be accepted by the awakening soul. In this light, the gap between much of today's popular spirituality and the mystery teachings that these belief systems imitate is hard to miss.

It is also important to understand, however, that these garbled echoes of the mystery teachings are not simple falsehoods. They have their own truth; it is just that these are partial and unbalanced truths. It is both true and important, for example, that many of us prevent ourselves from making the best use of the opportunities we encounter in life because our assumptions and attitudes get in the way. Convince yourself that you are certain to fail, and you will generally be right; convince yourself that you are certain to succeed, and the confidence and energy that come from that belief can lead to remarkable successes. Thus, the belief that you create your own reality can sometimes make it possible to accomplish things that more restrictive ways of thinking can prevent.

The difficulties creep in because the belief that you create your own reality is a partial truth, not a complete one. It needs to be balanced with its opposite, which is that your reality—the reality of the universe that was here before you came into being and that will be here long after you are gone—created you and continues to create you at each moment. Furthermore, the self and the rest of the universe join together at each moment to create the reality you will experience in the future. Thus, there are things you can accomplish and things that no amount of tinkering with attitudes will allow you to achieve, because they are contradicted by the momentum of greater processes in the universe as a whole. Equally, there are things that you might be able to achieve but should not attempt, because the achievement will set patterns in motion in the universe that will not be welcome to you.

These difficulties are best treated as an opportunity for learning. A century ago, in fact, many mystery schools taught their students, as a first exercise in exploring the power of thought, to believe that they create their own reality. The students would be assigned this and left to their own devices for a month or so. Very often they would return to their classes or write back to their tutors in a state of fair bewilderment, because some things that seemed almost impossible to them had happened easily, and other things that seemed easy had not happened at all. The teachers or tutors would smile and say, "Good. Now that we have your attention, we can start making sense of what the mind can do and what it cannot."

The book you are holding in your hands thus could just as well have been titled *Now That I Have Your Attention*. For those who have experimented with the power of the mind to change circumstances and are puzzled by their successes as well as their failures, the ideas covered in the chapters ahead offer a next step in understanding. In the process, this book inevitably challenges many of the popular teachings of our time and may prove upsetting to those who have become emotionally committed to those teachings. Still, that cannot be helped.

There are many ways to make sense of the teachings of the mysteries, some more relevant to the present age than others. Perhaps the most useful of all just now, as already suggested, is to trace the ecological laws that govern whole systems in nature, where they can be experienced directly with the senses. One of the core mystery teachings explains that the macrocosm (literally "the great universe," the universe around us) and the microcosm (the "little universe," the universe within us) are mirror images of one another.

Thus, we can look to the world of nature around us for help in understanding our own nature, recognizing that if a theory about the nature of the universe proves to be a mistake when tested against the world around us, it will also prove to be a mistake when applied to the world within us. As the great nineteenth-century teacher of the mysteries Eliphas Lévi put it, the core doctrine of the mystery teachings is that "the visible is for us the measure of the invisible."

THE SEVEN LAWS

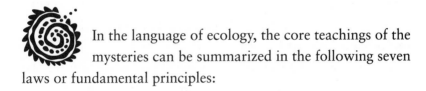In the language of ecology, the core teachings of the mysteries can be summarized in the following seven laws or fundamental principles:

The Law of Wholeness

The Law of Flow

The Law of Balance

The Law of Limits

The Law of Cause and Effect

The Law of the Planes

The Law of Evolution

In the pages that follow, we will consider these laws one at a time, using examples from nature as our touchstone. Following each commentary is a meditation, an affirmation, and a theme for reflection, to help bring the seven laws of spiritual ecology from the plane of abstract ideas into the realm of everyday life, where they become keys to meaning, wholeness, and power.

The First Law: The Law of Wholeness

Walk out into a grassy meadow, and you will find yourself surrounded by many different living things. Grasses and herbs of many kinds raise their leaves to the sunlight. Here a cricket crouches on a stem; there a bee visits a flower; a sudden flurry of motion shakes the grass as a field mouse falls victim to a snake; in the air overhead, a hawk circles lazily. Take a trowel and dig into the soil, and you will find earthworms, centipedes, and many other creatures going about their lives. Take out a hand lens, and a new world of life opens and becomes visible, populated by countless living things too small for the naked eye to see; view a sample from any part of the meadow through a microscope, and another dimension of the meadow's life, full of strange and wondrous living things, opens out before your eyes.

Each of the creatures you encounter, large or small, experiences its world in much the same way that you experience yours: as a background to its own actions—a background containing some things that are desirable, others that are dangerous, and still others that mean nothing to it and to which it pays little attention. To the field mouse, the important parts of the meadow are a burrow in which to hide, trails along which to scurry, other mice with which to interact according to the customs of mousekind, seeds and other edible things to eat, and snakes and hawks to avoid. To the snake that eats the mouse, the meadow consists likewise of homes and routes, creatures of the same kind, and creatures of different kinds, all against a background pattern of many other things that are not of interest to snakes.

The science of ecology, however, teaches that there is a different way to observe the meadow, one that reveals patterns that cannot be seen from a mouse-centered or snake-centered view—or, for that matter, a view that focuses on any other single species, such as humanity. From this ecological perspective, the meadow is a whole system, and all the living and nonliving things that exist there are parts of that whole system rather than wholly independent existences. The ecologist sees sunlight falling on the leaves of plants and recognizes that part of the energy in that sunlight passes from leaves to seed, from seed to mouse, from mouse to snake, and from snake to hawk in a single process. Thus, the meadow is more than the sum of its parts; it is a unity with its own rules, relationships, and patterns that are distinct from those of the individual beings that are part of that unity.

The meadow is a whole system, and all the living and nonliving things that exist there are parts of that system rather than wholly independent existences.

To observe the meadow as a whole system is to realize that each living thing in the meadow depends on all the other parts of the whole system for its survival. The field mouse, for example, does not get the food it eats, the water it drinks, or the oxygen it breathes out of nowhere; it gets them from the whole system of the meadow, or from one of the larger systems of which the meadow itself is only a part. Everything that the mouse gives back to the meadow, from its breath to its wastes to its body, is taken up by some other part of the whole system and put to some other use. Every part of the whole system is connected to every other part and can only be understood in its relation to the whole system of the meadow.

Look more closely at the field mouse, and the same perception can be applied to a new level. The mouse itself is a whole system composed of other, smaller entities—the cells that make up its body. Each of these cells is a little life of its own, which is born, lives for its own short lifespan, and dies. Each cell receives what it needs from the whole system we call the mouse, and it gives back its wastes and, eventually, itself to the whole system, to be put to some other use. Just as the field mouse is linked to every other living thing in the meadow, the cells that compose the mouse are linked to every other cell, and they can only be understood in their relation to the whole system of the mouse.

The dependence of the cells on the mouse and the mouse on the meadow is no abstract matter. If some of the cells in the mouse's body stop fulfilling their proper function in the whole system, the mouse becomes sick, and if the cells do not return to their function, the mouse may die. If the mouse dies, however, the cells die too; their life depends on the life of the whole system in which they are a small part.

The same is equally true on the scale of the meadow. If the field mice in a meadow stop fulfilling their proper function, the whole system of the meadow is damaged, and the mice suffer as a result; if the damage they inflict on the meadow is severe enough, they may all die. No part of a whole system can ignore the needs of that system with impunity.

As a principle of the mystery teachings, these same insights can be summed up in the language of ecology as the Law of Wholeness:

Everything that exists is part of a whole system and depends on the health of the whole system for its own existence. It thrives only if the whole system thrives, and it cannot harm the whole system without harming itself.

This principle applies to human beings just as fully as it applies to mice and meadows. Every human being participates in whole systems at many different levels, which nest within each other like the layers of an onion or the orbits of planets in a solar system. Some of these systems are local, some regional, and some planetary; some are made up of other human beings and their creations, and some are made up of other, nonhuman living things and the things that they produce. Our participation in these whole systems is not optional. Everything that makes our lives possible, joyful, and meaningful comes to us from them.

If we ignore our responsibility to the whole systems around us, the results are predictable—and unwelcome. What is the rising spiral of environmental crisis that now threatens to undo all the gains of the last three hundred years of industrial society, after all, but the inevitable result of our collective abuse and neglect of the living Earth over the course of those same three hundred years? The self-centered decisions that have put temporary profit and comfort ahead of the survival of whole species and ecosystems have had their inevitable effect. They have given us, and all other living things, a more impoverished and unstable world, and that impoverishment and instability has been reflected, in turn, in our own societies and our individual lives. The only way to undo the damage that has resulted from these shortsighted actions is to replace the habits that caused them with a recognition that, in all our actions, we need to put the enduring needs of the biosphere ahead of our immediate cravings for material comforts and pleasures.

This is why the sages and initiates of humanity have always taught that the way to happiness is found in living our lives in ways that benefit whole systems, rather than simply trying to benefit ourselves alone. All ethical teachings come down to this one point. It has become fashionable nowadays to laugh at such teachings, but we have seen what happens when individuals pursue their own benefit at the expense of the whole systems to which they belong. No matter how lofty the rhetoric used to justify them, actions that harm whole systems harm everyone and everything that is part of those systems, including those unwise individuals who think that they benefit from those actions.

If we try to get some benefit from the universe in a way that harms other beings, that harm will circle back to us through the whole system in which we all participate.

This is why it is a mistake to think that we have the exclusive power to create our own reality. We participate in the creation of our own reality, but we also help create the reality of every other being that has a place in the whole systems that surround us, and their attitudes and actions also help shape the reality we experience. Our attitudes and actions, in turn, can have effects we do not intend or expect. If we try to get some benefit from the universe in a way that harms other beings, that harm will circle back to us through the whole system in which we all participate, and whatever benefit we get will be balanced or outweighed by the harm that comes to us as participants in that system.

Putting this principle to work in our daily lives is a simple thing, though it is always worth remembering that "simple" is not the same thing as "easy." Nothing is so easy, or so self-defeating, as the habit of thought that claims that a decision affects nobody but the person who makes it, but this claim is never true. Even the most intensely personal of your decisions, made in the silence of your own heart and never expressed in any intentional action, radiates outward through voice, gesture, posture, and mood, as well as subtler expressions of consciousness, to affect the whole system around you. Choose to hate, even if you never act on your hatred, and you make the world around you a more hateful place, for you as well as for all other beings. Choose to love, and the reverse is equally true.

In making any decision, it is thus always important to think about how that decision will affect the whole system in which that decision takes place. That whole system may be a family, a workplace, a community, or a nation; it may include a local eco-system, a bioregion, or the entire planet. However large or small those systems may be, they will gain or lose by your decision, and that reality needs to be taken into account, for whatever gain or loss your decision brings to the whole system will circle around to affect your life as well.

The same principle can be put to use in a more deliberate way. If you wish to make your life better and happier, consider what actions you can do that will benefit the whole systems to which you belong, and do them. Those benefits will circle back to you through the connections that link you to the whole systems that support your existence. The relationships between the benefits you provide to the whole system and the benefits that return to you from the whole system are often subtle and complex, and they form the basis for many of the secrets of magic.

The essential lesson of this principle is that what benefits the whole system ultimately benefits the individual and what harms the whole system ultimately harms the individual. Thinking that an individual can pursue his or her own benefit at the whole system's expense is a folly that causes much of human suffering. The initiates of the mysteries recognize this; they seek first to understand how their choices influence the whole system and then take those influences into account when making decisions and acting on them.

Meditation on the First Law

The art of meditation is one of the fundamental practices of the mysteries, but mystery schools approach it in different ways. The most important of the differences is between those forms of meditation that empty the mind, practiced primarily in the mystery schools of the East, and those forms that guide and direct the mind, which are practiced primarily in the mystery schools of the West. The latter approach is the one that will be used here.

Each mystery school has its own recommendations for posture, breathing, and other aspects of meditative practice, but these are patterned with an eye toward the higher reaches of meditative experience, with which we are not concerned here. For the present purpose, you may use any reasonably comfortable position in which you can relax and remain still for a time. Settle yourself in your position, and then pay attention to your breathing. Breathe a little more slowly and deeply than you usually do, and as you breathe out, let go of some of the tension in your body, relaxing a little further with each outbreath. When you feel ready, turn your attention to the theme of the meditation, as follows.

Choose any material object that has a place in your life, and explore the ways that it connects with the world outside itself. Whatever it is, the matter of which it is made came from somewhere else and will go somewhere else; energies from some other source—whether those were the forces of nature, the skilled hands of a craftsperson, or something else—made it what it is now, and other energies will unmake it in due time; things that are not part of it provide the space in which it functions, the raw materials for its effects, and the meaning of whatever it does. All these are parts of the whole system of which that object is one small part.

As you work through this meditation, your thoughts will doubtless try to wander off onto different subjects. Bring them gently back to the theme of the meditation, and continue from where you began to wander. If any part of the meditation upsets you or makes your thoughts unusually eager to find something else to think about, make a note of that, and pay close and detailed attention to exactly that part of the process.

Repeat this meditation several times, choosing a different subject for each session of meditation. Thereafter, shift the focus of your meditations from strictly material objects to subtler realities, such as relationships and beliefs. Finally, do the same meditation with yourself as the subject at the center of the web of relationships and connections. Begin to become aware of the many whole systems of which you are one small part.

I am and all that I do, I am part of a greater whole."

How does the way you live affect the whole systems that support your life?

The Second Law: The Law of Flow

Walk out into the meadow again and look around yourself with eyes attuned to the Law of Wholeness, and you will be prepared to see another dimension of the whole system: a dimension of flow. The ray of sunlight that falls on a leaf sets in motion a flow of energy that passes from leaf to seed, to mouse, to snake, to hawk, shedding some of itself back into the environment at every step of the way. A drop of water that falls out of the sky as rain is picked up by the roots of a plant, rises through the stem to a leaf, transpires out of the leaf into the air, and rises up to the clouds to fall again as rain on some other meadow. Nitrogen gas from the air percolates down into the pores of the soil and then is taken up by nitrogen-fixing microbes clustered around the roots of a clover plant, turned to nitrates, absorbed by other plants that need nitrates to grow, and passed on from one living thing to another through many different changes of form until it finally rises again into the air.

To the ecologist, these flows are at least as real and important as the individual living and nonliving things that participate in them. Individual things come and go; the cricket on the stem will have been replaced by another in a year and by a different species of cricket in ten thousand years. Still, the flows that pass from the sun, the distant sea, the soil, and many other sources through the cricket that perches on the grass stem today will continue to flow through those crickets of the far future, just as they flowed through the crickets that came before it. In a deep and important sense, the cricket is simply a shape that is temporarily taken by flows of matter, energy, and information through one part of the whole system of the meadow.

Just as the relationship of parts to whole systems can be observed on many different scales of size, the relationships of flow can be observed on many different scales of time. Some flows are fast; for example, it may take only a few minutes for oxygen drawn into the lungs of a field mouse, transformed by its metabolism, and breathed out again as carbon dioxide to be absorbed by the pores of a nearby leaf, cycled through the plant's metabolism, and sent back out as oxygen again. Some flows are slow; for example, minerals absorbed by the roots of a clump of grass may find their way into its leaves, linger there through an entire growing season, then fall to the ground as the grass dies back at the beginning of winter and return to the soil with the reawakening of life in the spring. Some flows move at a pace so slow that human senses cannot perceive them at all; for example, the boulder left by a glacier on one corner of the meadow during the last ice age fifteen thousand years ago is slowly being weathered away by rain, wind, and the slow action of lichens, and fifteen thousand years from now, it will be a fraction of its present size. Solid as it seems, the stone is also flowing.

The rate at which flows move through a whole system can be changed by some of the things that parts of the system do with the flows that pass through them. The lichens on the boulder, by secreting chemicals that break down the rock to get at the minerals it contains, are speeding up the flow that is gradually turning the rock from part of a distant mountain into sand that will someday end up on the shores of a distant sea. The field mice who line their nests with dry grass, to provide themselves with warmth and food against the coming of winter, are slowing down the flow that normally turns minerals in the grass back into minerals in the soil.

Accumulation is increase without flow, and when it appears in a natural system, it shows that the flows in that system have begun to break down.

These small variations in speed, even if they happen at random, do not harm the whole system. If they happen in a predictable way, the whole system will adjust for them and may even come to depend on them. If the actions of part of a system distort the flows that pass through them drastically enough, however, the whole system can be damaged. The sign that this damage is happening is that accumulation appears somewhere in the system.

Accumulation is increase without flow, and when it appears in a natural system, it shows that the flows in that system have begun to break down. When grass seeds pile up on the ground uneaten, that is accumulation, and it shows that something has gone wrong with the field mice and other animals whose role in the meadow's ecology is to eat excess seeds. When the number of field mice increases beyond the level that the meadow can normally support, that is also accumulation, and it shows that something has gone wrong with the predators and other natural processes that keep the population of field mice at a stable level.

Accumulation is poison. Unchecked, it will destroy the system in which it takes place. As a result, whole systems respond to accumulation with responses that restore flow. If grass seeds pile up on the ground uneaten, they will attract other animals that eat grass seeds, and they will take over the role that once belonged to the field mice. If field mice become too abundant, the snakes and hawks that feed on them may also become more abundant, and they will turn the accumulation of mice back into flow by eating the excess. If something prevents the hawks and snakes from checking the field mouse population, the mice will eat up the food available to them and starve, and their bodies will return to the soil, restoring flow.

As a principle of the mystery teachings, these same insights can be summed up in the language of ecology as the Law of Flow:

Everything that exists is created and sustained by flows of matter, energy, and information that come from the whole system to which it belongs and that return to that whole system. Participating in these flows, without interfering with them, brings health and wholeness; blocking them, in an attempt to turn flows into accumulations, causes suffering and disruption to the whole system and all its parts.

This principle, like the Law of Wholeness, applies to human beings just as much as to mice and meadows. All the things that make it possible for humanity to thrive are flows that arise in the natural world, pass from natural systems to human beings, continue from one human being to another, and eventually flow back to the world of nature. Energy, for example, flows from the sun into leaves; nutrients flow from the soil into roots; information flows from the evolutionary past into genes of food plants. All these flows participate in the natural processes that put food on our tables. Energy, nutrients, and information also pass in various ways to the factory workers and machines that make the products we buy and use. The flows do not stop with us any more than they begin with us; our bodily wastes continue the flows that brought food to us, and what we do with our purchases after we are done with them determines what happens to the flows embodied in those products.

The contemporary spiral of environmental crises offers a useful lesson in the Law of Flow. Most people nowadays have forgotten that everything we receive from the whole system of nature flows from somewhere else, and everything that we put back into nature does the same thing. Most people have forgotten, as well, that the whole system of human society is also subject to the Law of Flow, so that everything we take has to come from somewhere and everything we put back into society goes somewhere and does something. When we say that something "goes away," we are speaking nonsense, for there is no such place as "away." Just as everything that comes to us comes from somewhere, everything that leaves our presence goes somewhere, and the place it goes is affected by the flow we set in motion.

In most human societies nowadays, a system of abstract tokens is used to help manage the flow of energy, material, and information from person to person in the form of goods and services. We call these tokens "money." There is nothing inherently wrong with money or with systems of exchanging goods and services among people that rely on money in order to function; we call these systems "economies." Problems arise, however, when people forget that money is simply a way of measuring and distributing the real wealth of goods and services, and they try to accumulate it rather than participating in its flow.

Just as lichens speed up the flow of nutrients by extracting them from rock and field mice slow down the flow of nutrients by gathering dry grass in their nests, human beings often need to speed up or slow down the flow of money. We may save money at one stage of life in order to be able to spend it in another, or we may use it to pay for an investment that will bring us needed returns later on. These are ways of participating constructively in the flow of money. Flow gives way to accumulation, however, and constructive participation changes into a destructive force when amassing money becomes an end in itself or when the amount of money being gathered up goes beyond anything a person, a family, or a human community actually needs.

Problems arise when people try to accumulate money rather than participating in its flow.

This is a crucial lesson just now, because most people in this phase of the world's history are infatuated with a mindless passion for accumulating money. The poor person dreams of having a million dollars; the rich person dreams of adding another million to the millions he or she already has. As a result, accumulations of money block the flow of real wealth and poison nearly every aspect of our collective lives.

In such an environment, it can be difficult to remember that money is simply a tool for managing the flow of real wealth among people, and that choosing to pursue accumulation rather than flow can cause stoppages in the making and sharing of real wealth. These stoppages make life more difficult for everyone in the whole system that is affected by accumulation. Today's media calls these events "economic crises"; the initiates of the mysteries recognize them as failures to understand the necessity of flow.

The practical application of the Law of Flow is straightforward enough, though it needs to be understood with some care, for the legacy of an older set of spiritual misunderstandings has to be taken into account. The mystery teachings of an earlier time, reacting to the extreme materialism of the cultures of that period, once put a great deal of stress on the unimportance of material wealth in the larger scheme of things. That emphasis was taken up and distorted by the popular spirituality of that age into a claim that material wealth was an evil thing and that spirituality and poverty were one and the same.

Like the belief that each of us creates our own reality, this is a partial truth rather than a simple falsehood. "Where your treasure is," a great initiate of the mysteries once said, "there your heart will be also," and the attitude that puts the pursuit of material wealth ahead of other human values is a serious obstacle to any form of spiritual development. One of the best ways to get past that obstacle is to deliberately embrace poverty for a time. Many of the old mystery schools, accordingly, had their pupils take up a radically simple, even impoverished, lifestyle for at least a certain period of their training, so that the students could learn just how little in the way of material wealth a human being actually needs to live a healthy and functional life.

As so often happens, that useful discipline was taken out of context in unhelpful ways. More recent presentations of the mystery teachings, which made the point that poverty was not a requirement of an enlightened state, were taken out of context in turn. This is how we ended up with a flurry of popular spiritual teachings claiming that an enlightened person ought to be surrounded by heaps of material wealth that he or she has essentially wished into being.

In all this confusion and misunderstanding, the Law of Flow offers a useful touchstone. If material wealth is flowing into your life, material wealth in some form should be flowing out of it at an equal rate. This does not mean that you have to spend every penny you earn; it means that you should plan on providing the whole system around you with value equal to the money you bring into your life. In more common language, it means that you should expect to earn the wealth you have and have precisely the wealth that you earn.

If material wealth is flowing into your life, material wealth in some form should be flowing out of it at an equal rate.

There are many ways to earn wealth, some obvious and some less so, but if you approach questions involving money from the point of view of earning it, rather than simply getting it, you will find that those questions are much easier to solve. Trying to get the universe to give you unearned wealth rarely works, and when it does work, it comes with a price tag that measures the cost you have imposed on other people, and on whole systems, by seeking accumulation rather than flow. If you ask the universe for a way to earn the money you need by creating something of value for others, on the other hand, you will normally find that the universe is more than willing to meet you halfway, or more than halfway, because you are seeking to participate more fully in the Law of Flow.

Meditation on the Second Law

The preliminary steps for this meditation are the same as those for the meditation on the Law of Wholeness. When you have settled into your position and spent several minutes focusing on your breath and letting yourself relax, turn your attention to the theme of this meditation, which is as follows.

As in the meditation on the first law, choose a material object that has a place in your life and learn to see it as part of a process of flow. Become aware of the whole trajectory of its existence and of the existence of the materials that compose it, from their origins in nature to their final destiny, whatever that happens to be. Pay attention to the subtle changes that affect even the most seemingly permanent and enduring things—the slow wearing away of substance and change of form that turns one thing into another. Pursue the meditation until you can see the subject of your meditation not as a thing, but as a process, something constantly changing and flowing along the trajectory of its existence, in response to the Law of Flow.

Do this meditation with several material things, devoting one session of meditation to each subject. Thereafter, turn your attention to less strictly material subjects, such as beliefs and relationships, and explore the way that these, too, flow and change throughout their lifespans. Finally, take yourself as the subject of this meditation and learn to experience yourself as a process of flow that involves matter, energy, and information, constantly changing as you flow from birth to death—or, as the mysteries teach, from long before the birth of the physical body to long after it has dissolved into the constant flow of material substance that brought it into being.

"What I receive from the universe is the measure of what I give to the universe."

Where in your life does accumulation keep you from benefiting from the Law of Flow?

The Third Law: The Law of Balance

Return to the meadow a third time, with the ideas of wholeness and flow in mind, and you will be ready to notice a third dimension in the dance of nature's patterns that surrounds you: a dimension of balance. The energy that descends in a ray of sunlight onto a leaf is balanced by the energy that is given back to the environment by all the lives that participate in that energy. The water that flows into the meadow as rain is balanced by the water that flows out of the meadow as water vapor and groundwater. The nitrogen that is drawn down from the air by the work of nitrogen-fixing bacteria is balanced by the nitrogen that is returned to the air by the work of other bacteria. If a whole system is thriving, you can be sure that the flows that pass through it are in balance.

The same rule of balance applies to the living things that participate in these flows. The number of field mice in the meadow is kept in balance by the amount of grass seed produced each year for mice to eat and the number of hawks and snakes in the meadow that feed on mice. Every living thing in the meadow is subject to similar balances and holds the balance, in turn, for some other living thing.

Look again at the field mice, and the way that this balance works is easy to understand. Every year, a new crop of grass seed falls to the ground; every year, new litters of infant mice are born; every year, new litters of infant snakes and broods of infant hawks hatch from their eggs. If the crop of grass seed is small, the litters of mice will be smaller, and the food supply of the snakes and hawks will be curtailed by this very fact. Fewer infant snakes and hawks will survive to prey on future generations of mice. If the crop of grass seed is large, the number of mice will increase, but so too will the number of snakes and hawks. Both processes keep the number of mice in the meadow relatively stable. The population of mice will rise and fall over the years, to be sure, moving to either side of the point of balance, but the balance of the whole system makes these shifts self-correcting.

Balance is not stillness, but a dance of constant adjustment around a common center.

If you examine the thermostat that keeps your home at a comfortable temperature, you will see the same process in action. If the house gets too cold, the thermostat turns on the furnace and heats it up; if the house gets too warm, the thermostat turns off the furnace and lets it cool down. No thermostat ever keeps a house exactly at a constant temperature, for the small shifts up and down the temperature scale are needed to make the system function. When a thermostat is set properly, though, these shifts never move the house outside the range at which the residents feel comfortable.

Like a thermostat, balance in whole systems always works in both directions, and it works by encouraging flow. When any part of a system begins to suffer from any form of accumulation, the system works to diminish the excess by converting it to flow and allowing it to move elsewhere in the system. When any part of a system suffers from a shortfall of anything, the system works to bring additional flow into that part of the system from elsewhere. Since every whole system participates in larger systems and receives what it needs from the endlessly changing flows in those systems, flows anywhere in a system will always be rising or falling as the system maintains its balance. Similarly, the living things that participate in a system will always be increasing or decreasing in numbers and activity as they adjust to the changing flows of the system. Balance is not stillness, but a dance of constant adjustment around a common center.

These adjustments can take place on many different scales of time. Some balancing processes in nature are so fast that, to human eyes, they seem instantaneous, and the shifts to one side and then to the other of the balance are so small as to be hardly noticeable. Other balancing processes unfold on longer time frames, and some of these time scales are so long that entire human generations can pass through their lives while the balance is swinging one way or the other.

Here, as in so many cases, the common human tendency to treat our own perceptions as a measure of the universe misleads us. A shift in the balance of a system so fast that we can barely detect it might well seem ponderously slow to microbes with a life span of a few hours, while the balancing processes that affect human communities and seem to take a long time from our perspective would appear to pass in a blur of speed from the perspective of a bristlecone pine. This is important to remember, for many of the most important balancing processes in the world at present move at a slower pace than human beings are used to perceiving, and very often human beings think that a movement in a particular direction is a permanent fact of existence when it is simply one slow arc of a balancing process that will reverse itself in its proper time.

As a principle of the mystery teachings, these same insights can be summed up in the language of ecology as the Law of Balance:

Everything that exists can continue to exist only by being in balance with itself, with other things, and with the whole system of which it is a part. That balance is not found by going to one extreme or the other or by remaining fixed at a static point; it is created by self-correcting movements to either side of a midpoint.

Like the first two principles we have examined, the Law of Balance also applies just as much to human beings as to mice, meadows, and everything else in the universe. Every dimension of human life is subject to balancing forces that make movement to either extreme cost more, in every sense of that phrase, than movement toward the middle.

This truth can be difficult for human beings to accept, because so many of us mistakenly think that the things we want out of life are the opposite of the things we don't want and that the middle point of the balance between these is an unsatisfactory mix of hopes fulfilled and unfulfilled. We think of wealth as the opposite of poverty, happiness as the opposite of sorrow, and so on, and so we flee from the center point of balance in the hope of finding some way to have whatever we happen to want all the time. This is understandable, but it is also misguided, because the good things in life are not the opposite of one unwelcome thing, but the midpoint between two.

Thirst, for example, is too little water, while the opposite of thirst is too much water. Right in the middle, between too little and too much, is an amount of water that is enough to satisfy without being excessive. People who feel thirsty do not generally decide that since drowning is the opposite of thirst, drowning must be as delightful as thirst is uncomfortable, and respond to feeling thirsty by drowning themselves in a lake. Surprisingly often, though, people apply the identical logic to other balances in life, where it works just as poorly.

Money is an excellent example. Surveys have discovered that people at every income level, from single mothers struggling to make ends meet on minimum wage jobs right up to Wall Street executives trying to find a profitable investment for a twenty million dollar annual bonus, are convinced that they would be happy if they only had twice as much money as they do. The single mothers are probably right, but the executives are certainly not, and the point where rising income stops adding to happiness—which has also been tracked by any number of surveys—is a great deal closer to the income level of the single mothers than it is to that of the executives.

A complex tangle of motives underlies this curious phenomenon. Very often, those who spend their lives chasing more money than they have actually desire something else, which they think they can only get by having plenty of money. They may desire a feeling of security, or happiness, or self-respect, or confidence, or some other emotional state, but they have confused this emotional state with the money they think they need to buy it or with material goods, which can be purchased with money, that they think will provide it.

They are quite wrong, because our emotional states are defined by the expectations, assumptions, and habits of thought that each of us absorb passively over the course of our lives or create actively by our own deliberate choices, not by the numbers printed on our bank statements each month. The way to achieve a feeling of security is to come to understand the mental habits that give us feelings of insecurity and to change those habits. Piling up money will do nothing to change those habits. Still, the person who is convinced that wealth is the only answer to feelings of insecurity too often simply decides that whatever amount of money he has at the moment isn't enough to do the job, and the hopeless pursuit goes on.

Self-knowledge, the heart of the mystery teachings, is the only way out of this trap. If we stop trying to make the universe cater to our unexamined desires and fears, and instead we take the time to understand what we actually want and why, it becomes easier to recognize that more is not better and too much is just as damaging as not enough. This realization makes it possible to use the Law of Balance consciously. This is done by recognizing the place where too much begins and moving back from it to the midpoint of balance, rather than trying to add excess to excess in the face of balancing forces that will inevitably win out sooner or later.

More is not better and too much is just as damaging as not enough.

There are also subtler ways of working with the Law of Balance—those that make use of the principle of the rebound. Deliberately push a balanced system one way, and you will make it swing back the other way with redoubled force. This principle is the rarely understood basis of the physical and emotional disciplines of asceticism that are so often part of training in the mystery schools, but that have often been pushed to extremes in popular religions.

Deliberately push a balanced system one way, and you will make it swing back the other way with redoubled force.

When a student of the mysteries fasts, for example, he or she is not under the illusion that food is bad or the belief that the body ought to be punished for wanting to eat. Rather, the acts of doing without meals for a moderate period of time and ignoring the physical body's cravings for food during that interval produce certain reliable effects that result from the Law of Balance. Once the fast is over, the digestion is improved and the enjoyment of food is increased. While the fast continues, on the other hand, it is easier for the mind to enter into certain states of clarity; furthermore, the strengthening of the will that comes from setting aside fixed habits and natural cravings remains with the student long after the fast is over.

The same benefits can be gained by setting aside the fulfillment of any other desire that is natural to human beings. None of these desires are bad or wrong; all of them have their place in a balanced human life. Refusing to express them for a certain time or in certain ways, however, makes use of the Law of Balance, freeing up energy that can be put to use in constructive ways when the balance swings back the other direction. Choosing not to express the sex drive for a time, or under some circumstances, can thus be used to bring increased physical vitality; choosing not to express the desire for ego gratification can be used to heighten self-awareness, and so on.

The principle of the rebound can be used with a fine degree of exactness. Decide what you want to achieve, and then deliberately go and experience the opposite state for a time; you will find it much easier to ride the movement of the balance toward your goal. This same principle is used unwittingly, in a destructive way, by those people who insist on trying to have some unbalanced state in their lives all the time and, thus, guarantee that the balancing rebound will bring them the opposite of that state. The Law of Balance affects initiates and uninitiated persons equally; the difference is solely that initiates know how to use the law to their advantage.

Meditation on the Third Law

Begin this meditation with the same posture and breathing you used to lead into the first two meditations. When you have done so, turn your attention to the theme of this meditation.

As with the first two meditations, choose a material thing that is part of your life to be the subject of your meditation, but this time explore the role of balance in that thing's existence and function. Start by considering what forces had to be in balance to bring the subject of your meditation into existence, and go on to notice what forces must remain in balance to keep it in existence. What would damage or destroy it, and what balancing factors prevent that from happening? What does it take in from the world around it, what does it give back to the world, and what holds these two in balance? Pursue the meditation until you have a clear sense of the ways that the subject is balanced between contending forces.

Do this with several material objects, devoting a session of meditation to each one, and then turn as before to considering things that are less material in nature. Finally, take yourself as a subject for meditation, and consider the many ways in which the Law of Balance expresses itself in your life.

AFFIRMATION

"Everything that happens in my life expresses the perfect balance of the universe."

THEME FOR REFLECTION

Where is the point of balance between not having enough and having too much?

The Fourth Law: The Law of Limits

Return to the meadow again, keeping the first three laws of spiritual ecology in mind, and you will be ready to become aware of a dimension of reality that surrounds and shapes them all: a dimension of limitation. Everything in the meadow exists within definite limits. Furthermore, everything in the meadow thrives not in spite of, but because of those limits.

Some of the limits that shape the lives in the meadow unfold from the Law of Wholeness, and they define the boundaries at which the actions of a living thing begin to disrupt the functions of the whole system that keeps that living thing alive. Some of those limits unfold from the Law of Flow, and they define the boundaries at which accumulation begins and necessary flows start to falter. Some of those limits unfold from the Law of Balance, and they define the boundaries at which movement toward an extreme runs up against balancing forces strong enough to push it back toward the center. All these are necessary results of the first three principles we have studied.

Yet a great many of the limits that define and support the lives of the mice and grass and other living things in the meadow do not come from the relationship between these things and the whole system that surrounds them, as the limits imposed by the first three principles do. They come from within the living things themselves; they represent the choices that make those living things what they are and that give those things the power and beauty they possess.

A field mouse, for example, has teeth and a digestive system that are fine-tuned to get nutrition from seeds and other concentrated plant foods, and so this is what field mice eat. They do not eat crickets, even though crickets are very nourishing; they leave crickets to the garter snakes. They do not eat herbs, even though herbs are very abundant; they leave herbs to the rabbits. They limit themselves to one kind of food, and as a result, their bodies and their behavior are exquisitely shaped to get and use that kind of food. Rather than jacks-of-all-trades, they are masters of one.

The sparrows that compete with field mice for grass seeds have accepted even harsher limitations to achieve even more spectacular results. Like most birds, they have embraced many limits and traded in a galaxy of other possibilities in exchange for the supreme achievement of flight. Every cell in their bodies is shaped by the demands of life in the air; their bones are as hollow as straws, their bodies stripped of every fraction of an ounce of unnecessary weight, from teeth to tailbones. Like the field mice, they are masters of a single trade, and its requirements are strict enough that every other option has dropped away.

In one way or another, this same principle of limitation is true of every living thing in the meadow, and it is true of the whole system of the meadow as well. Some of the limits at work in the meadow, as already mentioned, maintain the wholeness, the flow, and the balance of the meadow. Other limits are defined by the tradeoffs each living thing makes to fit itself for a particular niche in the meadow's ecology. Still others are imposed by the basic realities of existence. Only so much sunlight falls on the plants in the meadow, and only a certain fraction of that can be used for the everyday miracle of photosynthesis that keeps most of the Earth's living things fed. Only so much rain and snow falls to water the meadow and keep its plants green. As a result of these limits, in turn, the meadow can only support so much grass, so many field mice, and so on. Without these limits, in other words, the meadow would not be what it is.

Beauty is born when a flow of energy encounters firm limits, and the more perfect its acceptance of those limits, the greater the beauty will be.

All this may seem harsh at first glance, but the Law of Balance is at work in this context as well. It brings two immense benefits, already hinted at, in exchange for the limitations this fourth principle imposes.

The first of these benefits is that limitation is the source of beauty. Look at a blade of grass bending in the wind, forming an exquisite curve as it shapes itself to the moving air. Everything that makes the grass beautiful is a function of the limits that rise out of its own nature and its relationship to the whole system that surrounds it. The elegant lines of the blade have evolved to make the most economical use of limited energy and resources, for example, and the curve at which it bends measures the limit of the blade's strength in the presence of the wind. Remove the limits from the grass, and its beauty goes away. The same thing is true of all beauty, in nature as a whole and in the subset of nature we call human life: beauty is born when a flow of energy encounters firm limits, and the more perfect its acceptance of those limits, the greater the beauty will be.

This realization can be challenging to grasp. The second aspect of limitation is even more so, for it teaches that limitation is another name for power. To accomplish anything in a whole system—and "power" is simply a word for the capacity to accomplish something—it is necessary to accept the limits that are the other side of that accomplishment.

Power is born when a flow of energy encounters firm limits, and the more narrow the outlet left open by those limits, the greater the power will be.

The example of the sparrow already mentioned is a good illustration of this aspect. To achieve the power of flight, sparrows and most other birds accept strict and inflexible limits that prevent them from engaging in many activities that other living things can do. These limits are anything but arbitrary; rather, they are the other side of the power of flight itself, the things that have to be given up in order for that power to manifest. The same thing is true of all power, in nature as a whole and in that subset of nature we call human life: power is born when a flow of energy encounters firm limits, and the more narrow the outlet left open by those limits, the greater the power will be.

As a principle of the mystery teachings, these same insights can be summed up in the language of ecology as the Law of Limits:

Everything that exists is subject to limits arising from its own nature, the nature of the whole system of which it is a part, and the nature of existence itself. Those limits are as necessary as they are inescapable, and they provide the foundation for all the beauty and power each existing thing is capable of manifesting.

In the symbolism of the mysteries, out of every group of seven, the fourth is the center, the hinge on which the others turn. The Law of Limits functions as that central pivot of the seven laws of spiritual ecology, the key to understanding the entire structure of the cosmos as ecology and the mystery teachings both reveal it. It is also the least popular of the seven laws presented here. More precisely, it's the one that most offends the ingrained prejudices and unquestioned assumptions of our age.

Suggest nowadays that limits are real and not simply the product of false assumptions in the minds of the unenlightened, and you can count on a bumper crop of disagreement and even condemnation. Suggest that limits are not only real, but also the source of all the beauty and power in the world, and you will most likely find that your listeners have passed beyond disagreement to blank incomprehension. A great deal of modern popular spirituality treats the idea of limitation as something equivalent to humanity's original sin, the negative force that keeps human beings from experiencing some better existence. Meanwhile, the people who make these claims, like every other living being, rely on the Law of Limits at every single moment.

We can stand up, after all, because the hard limits of our skeletons give our muscles a solid basis to work against, and we can sit down because we can trust chairs to limit our bodies' propensity to fall toward the center of the Earth. Our immune systems put limits on the ability of microbes to make us sick, and our skin and tissues put equally definite limits on the movement of the bodily fluids that keep us alive. The sphincter muscles that keep the products of our digestive and urinary systems from flowing out of us moment by moment also impose limits worth remembering and valuing! Each individual cell in our bodies is able to live and function only because it can shelter within the protective limits of its cell wall, and when a cell ignores the limits placed on it by the body as a whole system and instead grows in an unlimited way, doctors call that condition "cancer." Freed from all limits, the human body would not become something superhuman; it would simply turn into a puddle of red slush, powerless, ugly, and dead.

It may be objected that while limits are important to physical bodies, things are different in the realm of the mind—that the human mind is infinite. That notion has become very popular in recent decades, but it is also nonsense; and it takes very little attention to experience to show that it is nonsense. Try, for example, to think through two complex trains of thought at the same time; try to keep a single image in mind, without your attention straying onto any other subject even for an instant, for twenty minutes at a stretch; try to remember, word for word, a newspaper article you read last month. A being with an infinite mind would be able do these things without the least difficulty. Can you?

What makes these examples particularly important is that, with appropriate training and much practice, it is possible for most people to learn to do the three things just named. Those who are unusually gifted can learn to travel further along the same lines, pursuing three distinct trains of thought at once, for example, or holding a state of perfect mental focus for hours at a time, or memorizing entire books word for word. None of these skills come easily, however, and there are upper limits to how far any one person can develop any of them.

There are also issues of diminishing returns and foregone opportunities to consider. Someone with a talent for memory might indeed be able to reach the point where he could read an entire newspaper once and then repeat every story back word for word, but the amount of time and effort needed to develop that ability would leave very little time for other, potentially more valuable activities. An infinite mind would not be limited in this way. This distinction shows the difference between an infinite mind and a finite one with capacities for improvement and shows also that human beings have the latter, not the former.

Our human bodies and minds, then, are finite, and so are the bodies and minds of all the other beings in the universe, from the smallest to the greatest. According to the teachings of the mysteries, this fact reflects a central spiritual truth. Nothing can enter into manifestation except by becoming finite and, thus, embracing limits. Limitation is the nature of manifestation.

Nothing can enter into manifestation except by becoming finite and, thus, embracing limits.

While something remains infinite and limitless, it remains unmanifest, and it cannot enter into manifestation except by embracing limits.

The art of manifestation, in fact, is the art of limitation—the art of knowing what to limit and how, so that the infinite flows into a specific, limited, finite form and brings the unmanifest into manifestation. This is what happens whenever a child is conceived: the countless possible combinations of the parents' genetic material settle out into a single set of chromosomes, so that eyes that could have been any color, for example, become a particular shade of blue. This is what happens when a painter stands at her easel and creates a painting: the possibilities are infinite until the first spot of paint touches the canvas, and each brushstroke narrows the remaining possibilities further until the painting is finished, fully manifest and fully finite. This is what happens, without exception, in every act of creation.

The practice of meditation, the core of the spiritual training offered by the mystery schools since before the beginning of history, shows the same principle in another way. From beginning to end, the art of meditation is an exercise in embracing limits. In meditation, the body is put in a position of balance and held there, or moved through a specific set of motions. The breath becomes slow, even, and silent; the mind is stilled or directed through some prescribed pattern of imagery or thought; every part of the self accepts strict limits for the period of daily meditation. Like the limits that enable birds to fly, these limits enable consciousness to turn back to its source and accomplish the work of awakening the higher dimensions of the self—the work that the mystery schools exist to teach.

More generally, the Law of Limits applies to every aspect of human life in the same way as the Laws of Wholeness, Flow, and Balance, and it can be put to practical use in the same way. If you want to achieve something in any part of life, the most powerful step toward achieving it is to figure out what limits you need to accept to make that happen and then embrace those limits. This is one of the great secrets of magic.

The Law of Limits means, among other things, that you cannot achieve two incompatible goals at the same time; you must choose one or the other. Many people fail to do this with money in particular. They want to be rich, but they also want to spend money as freely as they get it, and very often it never occurs to them that each of these goals prevents the other from manifesting. If you want to spend money freely, you can do that, but in that case, you might as well let go of your dreams of being a millionaire, because you will not achieve them.

The Law of Limits means that you cannot achieve two incompatible goals at the same time; you must choose one or the other.

If you want to become a millionaire, on the other hand, you can do that as well. All you have to do is get rid of every habit that prevents you from making a million dollars, and the habit of spending money freely is at the head of that list. Treat every dollar that comes into your hands as a tool to use in your quest for wealth and, thus, too precious to waste on anything else, and you will find ways to use each dollar to make yourself wealthier. Let making a million dollars become the most important thing in your life, elbowing aside every other aspiration, and you will infallibly make your million. Apply the same thinking to some less arbitrary and pointless goal, mind you, and it will work just as effectively.

Those who insist that the Law of Limits does not apply to them are still subject to it.

As the central pivot of the seven laws of spiritual ecology, the Law of Limits is one of the great secrets of the mysteries, and initiates use it constantly, choosing limits for themselves that will help them achieve power and grace in whatever work they are called to do. Those who insist that the Law of Limits does not apply to them are still subject to it; they simply lose the opportunity to choose their limits and to make their limits work for them rather than against them, and they struggle and suffer until they finally realize that there is a better way.

Meditation on the Fourth Law

Begin this meditation with the same posture and breathing you used to lead into the earlier meditations. When you have done so, turn your attention to the theme of this meditation.

As with the earlier meditations, choose a material thing that is part of your life as the subject of your meditation, and consider its limits. Make a mental list of the things it cannot do and cannot be. If some of the things on your list are absurd, this is not a mistake; it may seem silly, for example, to notice that a croquet ball cannot get up and walk away. Still, the croquet game in *Alice in Wonderland,* where the balls were hedgehogs who had the inconvenient habit of unrolling and wandering off on their own, may help you grasp one of the advantages of that particular limit!

When you have some sense of what the limits are that shape the subject you have chosen, go through the mental list you have made and think through the implications of each limit, seeing how that limit makes the subject of the meditation what it is, and how whatever beauty and power are in the subject unfold out of its limitations. It can be useful in many cases to see if you can think of ways to make the limits more strict and the subject's acceptance of those limits more complete, and thus increase the beauty and power of the resulting system.

Work through this process with several different material things, devoting a separate session of meditation to each one. Next, explore some of the less material realities in your life in the same way. Finally, apply the same meditation to yourself; consider the limits of your own life, body, and mind, and seek an understanding of the ways in which those limits make it possible for you to be what you are and to accomplish your dreams.

AFFIRMATION

"I embrace the limits that allow me to manifest beauty and power in my life."

THEME FOR REFLECTION

What in your life as it is right now would you be willing to give up, in order to have, do, or be the things you've always wanted?

The Fifth Law: The Law of Cause and Effect

Return to the meadow a fifth time, with the first four laws of spiritual ecology in mind. The Laws of Wholeness, Flow, and Balance establish the conditions under which all the living things in the meadow lead their lives, while the Law of Limits provides the solid foundation and pivot around which laws and living things alike function. In the interplay of these laws and lives, in turn, it becomes possible to perceive three more dimensions of the ecology of the meadow, and the first of these is a dimension of cause and effect.

If you pay attention to the happenings around you in the meadow, you will learn that nothing happens for no reason at all. Nothing happens merely by what human beings these days like to call "chance" or "coincidence." Instead, everything is part of a web of cause and effect that embraces every event, from the smallest to the greatest. Everything that takes place in the meadow is the combined effect of several causes and becomes, in turn, one of the

Everything is part of a web of cause and effect that embraces every event, from the smallest to the greatest.

causes of several effects. Some of the connections are obvious, and some are subtle; but an attentive watcher can follow the links of cause and effect from one end of the meadow to another.

Here is a little patch of bare dirt among the grass of the meadow. Why is it there? Look carefully to one side, and you will be able to catch sight of the mouth of the little tunnel from which the dirt came, the mouth of a field mouse's burrow. Why is the tunnel where it is and not somewhere else? Look again, a little past the tunnel's mouth, and you will find the remains of a rotting log that gives the mouse inside the burrow just that little bit of additional protection against predators. Why is the log there, in the middle of a meadow? Examine it carefully, using a pocketknife to scrape away the lichen that covers it, and you will find the blackened marks of a long-ago fire. You realize, looking around, that the meadow itself exists because, decades ago, a fire passed through the area and burned down the trees that once stood there. Look around again, and you may catch sight of the first few shoots of the forest that will stand there again, many years in the future, when the trees grow up and the meadow is no more.

Causes and effects always correspond to one another, both in scale and in kind.

Everything in nature finds its place, in much the same way, at the intersection of a complex web of cause and effect. Tracing the lines of this web, the attentive observer can very often read much of the past and at least something of the future. The key to understanding the flow of cause and effect, however, is to remember that causes and effects always correspond to one another, both in scale and in kind.

The correspondence of scale is easy to understand. The old saw about how small causes can lead to great effects—"and all for the want of a horseshoe nail"—is true in a way, but it must be understood in its proper context. The horseshoe nail that brought down the kingdom was not the only cause of that much greater effect. Rather, it was the tiny added weight that tipped a delicate balance one way instead of the other, when all the rest of the weight was already resting on the scales.

The forest fire that created the meadow, to return to the earlier example, may have begun with a single careless spark, but that spark would not have had the same effect without the right weather conditions, the slow buildup of dry, flammable matter over many years on the forest floor and among the shrubs and low branches of the forest understory, and a dozen other factors that set the stage for the fire. In the same way, the return of the forest begins with a seed blown in by the wind. Plenty of seeds, though, are blown to places where the conditions are not right for them to flourish, and many other factors must be in place before the meadow can become a forest once again.

More subtle, but equally important, is the correspondence in kind that must exist for a particular cause to lead to a particular effect. Even if all the other conditions for a forest fire are in place, the fire will not happen until a cause of the same kind as the fire—a discarded cigarette, say, or a lightning strike, or the sun's rays focused to incandescence through a glass bottle tossed aside by some thoughtless hiker—comes into play. A heated argument between two hikers, however inflammatory it might have been in metaphorical terms, would not have started the fire unless a physical spark resulted. In the same way, even if all the other conditions for the regrowth of a forest are in place, the forest will not begin to grow back until the actual seeds of forest trees arrive on the scene; no other seeds will do.

These examples are narrowly drawn, but the same principle applies to all things in the meadow and in the universe that surrounds it. Recognize the workings of these principles, and it becomes possible to trace effects back to their causes in the past and to follow causes forward to their effects in the future. In this way, given careful attention to issues of scale and kind, it can become possible to learn from the past and to shape causes in the present in order to produce specific effects in the future.

As a principle of the mystery teachings, these same insights can be summed up in the language of ecology as the Law of Cause and Effect:

Everything that exists is the effect of causes at work in the whole system of which each thing is a part, and everything becomes, in turn, the cause of effects elsewhere in the whole system. In these workings of cause and effect, there must always be a similarity of kind between an effect and at least one of its causes, just as there must be a similarity of scale between an effect and the sum total of its causes.

Very few concepts in today's world receive as much lip service as the idea of cause and effect, and even fewer concepts suffer from more misunderstandings. Most of these failures of understanding are motivated by a single factor: the very human desire, understandable but futile, to try to get something for nothing. That desire, and the pointless but passionate quests it so often inspires, can take many forms, from the blatantly materialistic to the seemingly spiritual, but they all come to grief on the simple fact that there is no way to achieve an effect without appropriate causes that are similar in scale and kind.

In other words, if you want something to happen, you need to figure out what causes would be adequate in scale and kind to make it happen, and then you need to put those causes into motion. Do this, and you can bring about wondrous changes in your life and the world around you. Leave this undone, and whatever happens, the results almost certainly will not be what you had in mind. They very likely will not be something that you wanted, either.

On the other hand, if you want something to happen, you also need to think about the consequences that will follow if the thing you want to happen actually happens, and you need to be sure you will be comfortable living with those consequences. It is for this reason that in some mystery schools the Law of Cause and Effect is called by a different name— the Law of Consequences.

The lesson taught by karma is that every action that a person does or leaves undone is a cause that has unavoidable effects.

The Law of Cause and Effect, in fact, is the law that shapes what the mystery schools of the East call karma. Like so many mystery teachings, this concept has been surrounded by a great deal of confusion over the years, but it is a very simple thing in reality. In Sanskrit, the sacred language of India, the word "karma" simply means "action." The lesson taught by the traditional lore of karma is that every action that a person does or leaves undone is a cause that has its unavoidable effects.

The same principle of cause and effect governs our relationship with money, that set of abstract tokens we use to help manage the flow of real wealth in our lives. The real wealth that money helps to manage consists of actual goods and services— on the one hand, natural goods and services, like fresh air, clean water, raw materials, and soil fit for growing food, which are produced by the labor of nonhuman beings, and on the other hand, manufactured goods and services, such as food, clothing, homes, and books like this one, which are produced by human labor out of natural goods. None of these things spring into being out of thin air, and none of them are limitless. The raw materials from which natural cycles and human labor produce goods and services are restricted by the limits of the finite Earth. The labor available from sunlight and rain, plants and animals, and human beings working at their trades is also limited by the laws that govern natural systems.

As the Law of Wholeness teaches, all these things come to us out of the wider world around us, and taking them for any given purpose affects the whole system of the biosphere; as the Law of Flow teaches, all these things are flows that must come from somewhere and go to somewhere else in the whole system; as the Law of Balance teaches, the movement of goods and services in one direction must be balanced by some equivalent movement in the other direction. Since money is simply a set of tokens that help us manage real wealth, human beings have learned that it makes sense to tie earning money to producing real wealth. You can't simply print out reams of money on your computer printer, for example, and if you could, money would very quickly become worthless and fail to do its job of helping us manage real wealth.

The way money works when understood as a whole system, when that system works as it should, is as simple as it is elegant. If you want money, you need to produce some form of real wealth that people want and will take in exchange for money. You can do this by being able to perform some kind of labor that an employer wants enough to exchange for money, or you can do it by being able to produce some good or perform some service that people who aren't employers want enough to exchange for money. As long as the tokens we call money work in this way, the system works. Once people start finding ways to get money without producing real wealth in exchange for it, however—and this happens to most money systems sooner or later—the system begins to break down, and money eventually stops being worth anything at all.

The unraveling of the money system of our society has been under way for many years now, and the economic troubles of recent years are the product of the process just outlined. Many people nowadays have spent their entire lives finding ways to game the economic system so they can accumulate money without producing any real wealth, and the undeniable fact that some of them have succeeded over the short term has probably done much to feed the belief that wealth is exempt from the Law of Cause and Effect—and even from the laws of common sense. The other side of their success, however, is the gradual breakdown of the money system itself, leading to the rising spiral of economic crises that is a common feature of the twilight of every civilization.

Those crises, in turn, have helped drive many people's desperate attempts to use spiritual methods to get the money they think they need, in order to maintain a standard of living that their place in a disintegrating economy will no longer support. Understandable as this attempt is, it is doomed to fail, because it feeds the problem to which it attempts to respond. It tries to attract real wealth without creating real wealth, and so it falls afoul of the Law of Cause and Effect. Those who choose instead to increase their own ability to produce real wealth through their own labor—if necessary, outside the system of tokens we call the money economy—may have less money but more real wealth and can generally count on a better outcome.

All this is simply the Law of Cause and Effect at work, in the specific context of the production and distribution of wealth. To create the effect of having real wealth in your life, you need causes of the appropriate scale and kind; in the long run, nothing else will do. Initiates of the mysteries understand this, and they measure the causes they set in motion according to their real needs; because they do, they are rarely rich, but always have what they need.

Meditation on the Fifth Law

Begin this meditation with the same posture and breathing you used to lead into the first four meditations. When you have done so, turn your attention to the theme of this meditation.

As with the earlier meditations presented in this book, choose a material thing that is part of your life to be the subject of your meditation. In this meditation, trace out the causes that brought that thing into being and then brought it into your life. If the subject of your meditation is a natural object, what caused it to come into being, and what caused you to give it a place in your life? If it is something created by human labor, what caused its maker to make it, its seller to buy and market it, and you to purchase it? Seek to become aware of the entire web of causes that lies behind every object, even the smallest and simplest.

When you have sorted out the causes of the subject of your meditation, turn your attention in the other direction and trace out its effects. What differences has having that thing made in your life, in other lives, and in the universe as a whole? What would be different if you had not chosen to bring it into your life?

Work through this process with several different material things, devoting a separate session of meditation to each one. Next, explore some of the less material realities in your life in the same way. Finally, apply the same meditation to yourself; consider the very complex web of cause and effect, including your actions and those of others, that brought you into being and made you the kind of person you are. Then consider what the effects of your present life will be on yourself, other people and living things, and the universe as a whole.

AFFIRMATION

"Everything I experience is the appropriate effect of causes in the past, and everything I do will be the cause of appropriate effects in the future."

THEME FOR REFLECTION

To what extent do your own actions cause effects you experience in your life, and to what extent do they cause effects that other people experience in their lives?

The Sixth Law: The Law of the Planes

Return to the meadow a sixth time and look around you with eyes attuned to the five laws of spiritual ecology that you already know. You will need to pay very close attention this time, for the dimension of the meadow you are trying to see is a subtle one: a dimension of distinct planes, levels, or modes of being that are present throughout the meadow and shape all the lives that exist there.

Start with the simplest things you see, and you will find it easier to achieve the awareness needed to grasp this dimension. Where a field mouse's burrowing has pushed a small mound of earth up onto the surface, you notice small stones and soil. The minerals that compose the stones and soil are found through-out the meadow, and some of them play important roles in the lives of the plants and animals there. Each mineral has its own place, but they all follow the same general laws; understand how potassium cycles through the meadow, for example, and you will have little trouble understanding the equivalent cycle that calcium follows.

Water also moves throughout the meadow, as humidity in the air, moisture in the soil, sap in the plants, and blood in the animals. It follows many of the same laws as the minerals, but not all of them, and it follows its own distinctive laws as well. Nitrogen and other gases, similarly, share some laws with minerals and water, but not all of them. There is a continuum of material substances, in fact, from the most dense and resistant mineral to the most fleeting and elusive gas, and all the material substances that participate in the whole system of the meadow range themselves along that continuum, following laws appropriate to their nature.

The step from matter to energy marks a break in the continuum. Everything in the meadow participates in the flow of energy, just as much as it does in the flow of matter, but energy forms its own continuum, embracing four primary forces—the atomic strong force and weak force, electromagnetism, and gravity—in an immense range of forms and intensities. It also follows its own rules. Try to make sense of energy using the laws that govern matter, and you can count on getting things wrong nearly every time, because matter and energy are different. In the language used in the mystery schools, they belong to different planes.

Does this mean that matter and energy exist in entirely separate realms? Of course not. Both share space in the whole system of the meadow and in the physical universe as a whole. Nor does any barrier prevent them from influencing each other; matter constantly affects the flows and patterns of energy, and energy just as constantly affects matter. You can even turn one into the other, though the conversion involves almost unimaginable amounts of energy and very little matter—around 10,000,000,000 kilowatt-hours of energy to a pound of matter, according to Einstein's famous equation, $E=mc^2$. This is why turning matter into energy involves nuclear reactions, with all their attendant dangers, and why turning energy to matter is something that scientists can only do one subatomic particle at a time, using enough energy in the process to light a medium-sized town. These huge differences are reminders that the line between two planes does not need to be absolute to be significant.

Turn your attention from matter and energy to life, and once again you have crossed a boundary between planes. Living things have bodies made up of matter, to be sure, and require constant flows of energy to survive, but their behavior cannot be predicted from the physical laws governing matter and energy. You may believe, as most scientists do nowadays, that life is simply a set of strange properties that emerge from complex combinations of matter and energy under the right conditions, or you may believe, as the traditional wisdom of the mystery schools has always taught, that life is a force that is distinct from matter and energy, a force that indwells living things and gives them their special properties. Either way, you cannot take a chemistry textbook and use it to understand the biology, life cycle, and behavior of a rabbit.

Life, in turn, shares space in the whole systems of the meadow and the universe with matter and energy. It affects them, and they affect it. It is possible that someday scientists will learn how to turn nonliving matter and energy into life, though they have not yet even come close to figuring out the trick yet; for the time being, as far as we know, life creates life, and only life can do so. Like energy and matter, life also has its own continuum, a spectrum of living forms encompassing everything from the tiniest microbes to giant sequoias and great whales.

Are there other planes beyond these? The mystery teachings have always held that there are, though it is true enough that without the special training the mystery schools offer, you will have a harder time seeing them around you as you stand in the meadow and watch the wind in the grass. The flow of information through the whole system of the meadow is one expression of another plane that has its own laws and its own spectrum of manifestation, but this plane is very nearly the only one that can be traced without using inner capacities that most human beings in the present age of the world have not developed.

Still, the basic point should be clear by now. The things that make up the whole system of the meadow do not all belong to the same mode of being or, in the terms used already, do not exist and function on the same plane. The distinction between one plane and another does not prevent things on different planes from existing in the same place and time and affecting one another—in fact, fairly often phenomena on one plane become evident only when something on another plane is present. Nonetheless, the different planes are different, and trying to treat things of one plane as though they are things of a different plane, or as though they can readily pass from plane to plane, is usually a mistake—sometimes a very large mistake.

The distinction between one plane and another does not prevent things on different planes from existing in the same place and time and affecting one another.

As a principle of the mystery teachings, these same insights can be summed up in the language of ecology as the Law of the Planes:

Everything in existence exists and functions on one of several planes of being or is composed of things from more than one plane acting together as a whole system. These planes are discrete, not continuous, and the passage of influences from one plane to another can take place only under conditions defined by the relationship of the planes involved.

Human beings are whole systems composed of things from several different planes, and different aspects of each individual exist and function on different planes. The teachings of the different mystery schools name and describe the planes in varied ways, for reasons rooted in the requirements of spiritual practice and the need to speak to each human culture in terms that it can understand. Many mystery schools speak of seven planes, and they divide each plane into seven subplanes; but not all define these planes in the same ways, and some sort out the planes into different arrangements entirely.

Exactly how the planes are arranged, however, is not important to our present theme. What needs to be grasped are the paired recognitions that, first, human beings are composite beings, existing on several planes at once, and second, the Law of the Planes, therefore, affects each one of us at every single moment.

Consider the plane of matter and the plane of mind. The teachings of every mystery school trace out a division between planes, separating the realm where thoughts exist from the realm where physical objects exist; so does modern science, and so, in turn, does ordinary common sense. There are connections between these two planes, to be sure. Every time you turn a page of this book, for example, a thought in your mind crosses from one plane to

Human beings are composite beings, existing on several planes at once.

another and causes the physical matter of your hand to move. Still, try to turn the page by an effort of your mind alone, without moving any part of your body, and the distinction between the planes becomes a little more important.

This is a trivial example, but there are many others that are less trivial. If you try to make changes in the world as you experience it by changing your thinking, you will find that some things change dramatically, others prove completely resistant to change, and still others fall in between. Why? Because the world that each one of us experiences is brought into being by things that exist on several different planes. Some of them are on the plane of mind and can be changed very easily by changing the way we think. Others are on planes that are close to the plane of mind—for instance, the plane of the emotions—and can be changed with a little more effort, by finding the point of contact between the planes and using changes on the plane of mind to set in motion corresponding changes on the emotional plane. Still others are on planes relatively distant from the plane of mind—for example, the plane of matter—and it's rarely possible to cause any kind of change there without bringing some more material force into play to assist the change in thinking.

The world that each one of us experiences is brought into being by things that exist on several different planes.

This is what confuses many people who try to make changes in their health or their economic status by changing their thinking without paying attention to the distinction between planes. Very often some things change for the better very quickly, since it's quite common for thinking to have direct or indirect effects on both of these. If you have a health problem that is caused by stress, and you change the attitudes that burden you with stress, the health problem can go away completely in a very short time. If your financial troubles are caused by attitudes that blind you to opportunities or lock you into the habit of spending more than you make, equally, a change in attitudes can transform your economic situation overnight.

Not all health problems are caused by thinking, however, and not all economic troubles are the result of attitudes. Those that have material causes will prove resistant to change when the change is pursued on the mental level, and they have to be addressed on the material level if they are to be changed. The same logic, it bears remembering, works equally well the other way: if the thing you need to change is on the mental level, changing things on the physical level is not going to do the trick. The countless people who try to feel better about themselves by packing their lives with consumer products and feel just as miserable as before are caught in this trap. Until they pursue change on the plane of the emotions, their efforts will yield them no benefit.

These mistakes happen because the Law of the Planes also applies to our perceptions of the universe around us. An earlier section of this book mentioned the problems faced by people who think they want one thing out of life, such as money, but actually want another, such as a feeling of security. This is ultimately a mistake in perception: a reality existing on one plane has been mistaken for a reality on another.

Such mistakes are particularly common nowadays because our contemporary culture teaches us some very bad habits when it comes to sorting out the perceptions that surround us. Most often, a child growing up will be told over and over again that only the things he or she experiences with the physical senses are real, while those things experienced with the other human senses are unreal. This teaching makes it very hard for the child to recognize that something he or she experiences with the emotional sense, for example, is as real as a rock on its own plane.

Still, those last four words—"on its own plane"—must be remembered. An emotion is as real as a rock, but it is not a rock. Emotional energy, powerful as it is, can influence the world of rocks only by passing through one of a small number of potential connections between the plane of emotions and the plane of matter. Nearly all of those connections are found in beings that exist on both planes, and the connections most readily available to human beings are in themselves—human beings, after all, combine emotional and material realities, as well as realities of several other planes—and in the people around them.

An awareness of these connections can sometimes allow for surprising effects—effects of a kind that many people consider "magical." The proven power of the mind and the emotions to influence the health of the physical body, for example, demonstrates that working from plane to plane in the right way can have remarkable results. Still, the fact that power can pass from plane to plane does not prove that the plane of matter is imaginary or that the mind's power over it is limitless. It simply shows that the individual human being is a whole system made up of material from several planes, and he or she participates in the universe as a whole.

Thus, the teachings of the mysteries focus much of their instruction on the connections between the planes found within each of us, so that the initiate knows how to direct an intention from one plane to another in the most readily available way and how to manifest it on the plane where it can become an appropriate cause for the desired effect. This training enables initiates to direct their energy where it will accomplish what needs to be done, instead of applying effort to one plane in an attempt to accomplish what can only be done effectively from another.

Meditation on the Sixth Law

Begin this meditation with the same posture and breathing you used to lead into the earlier meditations. When you have done so, turn your attention to the theme of this meditation.

As with the earlier meditations presented in this book, choose a material thing that is part of your life to be the subject of your meditation. In this practice, your goal is to understand the different planes of being present in that object. Begin by considering the material substances that are part of the item, and think about the roles those substances play in it. Next, consider the role of energy in the item and, after that, life—for most of the things around us are made at least partly from the bodies of living things, or they have been shaped by the actions of living things. From life move to the plane of the emotions, considering what feelings have shaped and been shaped by the item, and then go from the plane of the emotions to that of the mind, considering what thoughts have affected the item and been affected by it. Can you go further, to planes above thought?

Work through this process with several different material things, devoting a separate session of meditation to each one. Next, explore some of the less material realities in your life in the same way. Finally, apply the same meditation to yourself; seek an understanding of the roles that different forms of matter, energy, and information play in yourself, and then go on to explore other planes of yourself, including those of emotions and thoughts. Make this exploration as detailed and complete as you can.

"I respond to each reality I encounter on its appropriate plane."

THEME FOR REFLECTION

Which of the planes of the world you experience is most important to you?

The Seventh Law: The Law of Evolution

Return to the meadow one final time, and look around with eyes attuned to the first six laws. The dimension of the meadow that you are trying to see this time requires even more subtlety and patience to perceive than the ones we have already explored. You can grasp it, though, if you remember that everything you see around you, while it is what it is at this moment, is also a point along a trajectory defined by the movement of time. By paying attention to that movement and the changes it brings to every living thing, you can grasp the seventh and most secret dimension of the meadow: a dimension of evolution.

Consider a field mouse as it crouches next to a tussock of grass. That mouse stands at some point in its own personal life cycle, between the moment when it first huddled, blind and hairless, against its mother's belly, and the moment when it will meet its end under the teeth of a snake or the claws of an owl. The grass likewise stands somewhere along the arc of its own transformations, from the sprouting seed that brought it into being to the gentle efforts of the worms and bacteria that will transform its last remains into soil.

The mouse and the grass also stand at a specific point along the arc of a much greater transformation: one that will turn the meadow, over the course of several hundred years, from the burnt bare ground that followed a forest fire of the past to old-growth forest sometime in the far future. Ecologists call this process "succession," and it plays the same role in the history of ecosystems that the life cycle plays in the lives of individual creatures. Just as every creature begins with a single cell and passes through its life cycle, every ecosystem begins with bare, nonliving elements and passes through stages, called "seres." Those stages reach from the first or pioneer sere that forms on bare ground right up to the final relatively stable sere, which ecologists call "the climax community."

In temperate regions on land, the pioneer sere normally consists of fast-growing weeds and the insects and other small creatures that live on and around them, and the climax community is old-growth forest. It can take hundreds of years for a piece of ground cleared by a forest fire or human logging to pass through the whole sequence of seres and reach its climax sere, but unless some outside force interferes with the natural unfolding of succession, sere will follow sere in a predictable order until the climax community establishes itself. The meadow in which the mouse and the grass thrive, in other words, is simply one phase of a greater process of change that began long before either one was born, and that meadow will continue long after both have died.

The mouse and the grass, furthermore, also have a place along an even vaster trajectory. Ten million years ago, some other species of small rodent filled the same role in meadow ecologies in the same region that the field mouse fills today, and ten million years from now, today's field mouse will likely be replaced by yet another species of rodent, or some other creature not too different. The grass, likewise, fills a particular role—an ecological niche—in the ecosystem, one that was filled by other plants in the distant past and will be filled by still other plants in the distant future. The process by which one species gives way to another is evolution in the strict sense of the word.

Few ideas in all of human history have been more thoroughly misunderstood than the simple concept of evolution. Intellectuals in Victorian England, eager to use the science of their time to bolster a class system already cracking under the weight of its own injustice, invented the notion that some living things—and thus some people—are "more evolved" than others. That turn of phrase is still much used today, but in the real world, it is quite simply nonsense.

Every living thing is just as evolved as every other, because every living thing has been shaped by evolution over exactly the same period of time since life first evolved. Those creatures that have not changed for a very long time have simply adapted so well to their environments that the process of evolution maintains their current form rather than driving them onward to a new one. One could just as well say that those organisms that have not changed in half a billion years are so highly evolved that there has been no need for them to change.

Every living thing is just as evolved as every other.

Another set of intellectuals, a little later on, promoted the idea of "evolutionary leaps" to widen the distance between human beings and other living things and also between some human beings and others. In the real world, this is just as much nonsense as talk about "more evolved" and "less evolved" species. Seen from the perspective of geological time, evolution sometimes does seem to move quickly, but that word "quickly" needs to be measured against time scales almost unimaginable to human beings. The scientists who have pointed to variations in the rate of evolution and the tendency of species to remain in a stable balance for long ages and then change quickly when conditions change are still talking about millions of years. Within any human sense of time, evolution is an immensely slow pro-

Within any human sense of time, evolution is an immensely slow process.

cess, unfolding over countless lifetimes as the pressures of environment make certain traits more helpful than others.

Most of the time, on its own vast scale, evolution resembles nothing so much as the life cycle of an individual living thing or the process of succession in an ecosystem. It happens now and again that members of a certain species of living things find themselves having to face an unexpected environmental challenge, because some accident has opened a route that allows them to spread into new territory, because climate change has made what was once familiar ground different, or for any of a thousand other reasons. If at least a few of those living things are able to respond effectively to the challenge, they then find themselves under pressure to adapt in unfamiliar ways, and those of their offspring who are well suited to those adaptations will thrive in ways that the others do not. Let this process continue for many generations, and the group of living things will have evolved into something new.

If the adaptations prove to be stable, a new species is born and reintegrates itself into the whole system surrounding it, settling down to live in its new way for as long as environmental conditions remain the same. Just as the individual living thing goes through many changes in youth and then settles down to a stable condition in maturity, and an ecosystem goes through many seres in the succession process before it reaches relative stability as a climax community, each species moves toward a steady state and maintains that state as long as possible.

The world we live in is an unstable place, of course, and so any steady state will eventually be overturned by outside forces. When this happens, those living things that survive the challenge of the transition will adapt in new ways, new species will be born, and a new steady state will gradually come into being. The rhythm of challenge, response, and reintegration can thus occur repeatedly with a single individual, ecosystem, or species.

The core insight of this vision, however, is that evolution does not move in a straight line. Nor does it have a goal or purpose, other than adapting living things to the changing environments in which they live. Nor, crucially, does it make today's living things automatically better than the ones that came before them. The fox who hunts field mice in the meadow today is no better a hunter than the tritemnodon, an early mammal that hunted prehistoric rodents in the same place fifty million years earlier, or the saurornithoides, a fox-sized dinosaur that hunted even smaller dinosaurs in the same place fifty million years before that. Each of these predators fills the same ecological niche with much the same success, and when the fox goes extinct due to changes in the environment that go beyond its capacity to adapt, some new living thing will evolve to fill the empty niche with roughly the same effectiveness.

The rhythm of challenge, response, and reintegration can occur repeatedly with a single individual, ecosystem, or species.

Though evolution does not follow a straight line of progress, it can feel its way into new possibilities. Useful adaptations, once made, can enable living things to spread outward into new environments, adding their own adaptations to the mix, and the emergence and combination of these adaptations enrich the range of potential present in the whole system. Increased diversity, rather than progress along some preestablished line of development, is the keynote of evolution in the real world.

There is nothing certain in this process of expanding possibility, however. Whole ranges of adaptation that might have made the world richer for eons have been lost when the biosphere has shifted decisively and the living things that happened to bear those adaptations at that time failed to survive. Still, it is through the slow buildup of new adaptations that life on Earth has made the journey from a thin layer of single-celled organisms clinging to wet rocks some two billion years ago to the relative richness and complexity of the mature biosphere we have had for the last forty million years or so.

Notice, however, what does not happen in the course of evolution. No mouse ever made a sudden evolutionary leap to become a supermouse that was invulnerable to predators and could get its food from thin air. Neither the saurornithoides nor the tritemnodon ever leapt beyond the limits of the biosphere to which they each belonged, and the fox, clever and elegant though it is, shows no signs of doing anything of the kind, either.

The point to recognize here is that evolution is not a ticket to some imaginary place beyond limits and laws. Rather, like every process of creation, it transcends some limits by accepting others, gains certain goals by giving up other possibilities, achieves beauty and power through a harmonious relationship with its external and internal limits, and responds to the conditions set by the whole system in which it takes place.

As a principle of the mystery teachings, these same insights can be summed up in the language of ecology as the Law of Evolution:

Everything that exists comes into being by a process of evolution. That process starts with adaptation to changing conditions and ends with the establishment of a steady state of balance with its surroundings, following a threefold rhythm of challenge, response, and reintegration. Evolution is gradual rather than sudden, and it works by increasing diversity and accumulating possibilities, rather than following a predetermined line of development.

Like every other living thing, human beings have come into existence through a process of evolution that can be traced all the way back to the first, forgotten living organisms of the Earth's primal oceans some two billion years ago. Our bodies show the heritage of our long evolutionary descent in every detail.

We have, after all, the forward-looking eyes and nimble hands that the first primates evolved to deal with the challenges of living in the dense forests of the Eocene epoch 50 million years ago. We have the hair that the earliest mammals evolved from scales to deal with the harsh winters of the Permian period 250 million years ago. The basic plan of our bodies, with four five-toed limbs extending in pairs from a central trunk, can be traced back to the first archaic lungfish that crawled ashore from the warm seas of the Devonian period 400 million years ago. Even the chemical composition of our blood mirrors that of the oceans of the Archean era more than 600 million years ago, when the first living things large enough to need blood circulation evolved ways to pump a close equivalent of the water of forgotten seas through their tissues.

Do our minds and spirits show signs of the same trajectory? The teachings of most mystery schools, as well as those of modern evolutionary biology, agree that they do. Human thought differs from the thinking of most other animals in that we are capable of thinking in abstractions—not just this piece of food or that dangerous animal, for example, but also the abstract idea of food in general or danger in general. Combine this gift for abstraction with the nimble hands we got from our primate ancestors and share with our primate relatives, and you have an animal that can do things with the material world that no other animal can manage.

It is common enough for human beings these days to stretch this particular bit of uniqueness much farther than it will reach and insist that humanity is set completely apart from the rest of the animal world. This conclusion hardly follows. Every living thing has its unique gifts, and our combination of abstract thinking and physical manipulation seems so impressive to us principally because it is ours.

Nor are we unique in our ability to think in abstractions. All the evidence suggests, for example, that dolphins and whales are just as capable of abstract thought as you and I. The sounds they make to communicate to one another are as rich, complex, and changeable as any human language, and their brains are at least as complex as ours. Flippers cannot make or use tools, however, and so the cultures of dolphins and whales have evolved in different directions. Given our human propensity to destroy one another with the works of our hands, a case could easily be made that their evolutionary direction has turned out to be the wiser and happier one.

As for the realm of the spirit, most mystery schools teach that each human soul has made the same gradual evolutionary journey from simplicity to complexity that we find reflected in the human body. Our souls are not the only ones that have made such a journey, according to these teachings; in fact, these same teachings say that every living thing—and much of what we like to call "dead matter"—is a form taken by souls engaged in the same slow process of development that has shaped each of us and will continue to shape us far into the future.

Each human soul has made the same gradual evolutionary journey from simplicity to complexity that we find reflected in the human body

Central to the mystery teachings, too, is the understanding that the keynote of spiritual evolution, like that of physical evolution, is the increase in diversity over time. The great sages, saints, and initiates of our species are unique personalities; it is no accident that in one branch of the Western mysteries, the term used for the highest grade of initiation—*ipsissimus*—is a Latin word meaning, literally, "most completely oneself." Once again, we are part of a wider world, not exceptions to it.

The adaptations that took a species of primates closely related to chimpanzees, reshaped them to cope with the challenges of the harsh environment of the East African savannas, taught their hands to use sharp rocks and pointed sticks against predators and each other with equal skill, and transformed their hooting and chattering into abstract languages sent us in a unique direction. Like every other evolutionary process, the trajectory of our origins took us through many changes early on and then gradually settled on a common pattern with local and regional variations, the mature form or climax community of our kind.

It is a total misunderstanding of evolution to insist that we can expect it to solve all our problems in the near future, or at all.

That common pattern is still capable of change in response to changing circumstances, of course, and the slightly more abstract version of monkey-chatter that humans use to think can pass through its own rhythm of challenge, response, and reintegration much more quickly than our genetic inheritance; this flexibility is among the reasons our species has been successful, at least in the short term. Still, it is a total misunderstanding of evolution to insist that we can expect it to solve all our problems in the near future, or at all.

From a wry but more accurate perspective, in fact, evolution does not solve problems—it causes them. When the primates who became our first hominid ancestors left the trees for good maybe five million years ago in eastern Africa, after all, they were not escaping from some dire crisis in the African forests to embrace some higher way of life on the open savannas. Quite the contrary; the climate changes that were moving the Earth toward its most recent round of ice ages drove those primates out of forest homes, to which they had become perfectly adapted by millions of years of evolution, and forced them to find ways to survive in the far more challenging and dangerous environment of East Africa's plains.

Forced to seek unfamiliar foods and defend themselves against leopards and lions on open ground, the early hominids had to scramble for survival, and the adaptations that succeeded in giving them an edge in that desperate struggle gradually turned them into us. Their problems eased only when the process of human evolution was complete—when hominids had become humans, well adapted in body, mind, and spirit to a life that had challenged their ancestors to the core, and equipped with languages, social structures, and tool-using skills that gave them a comfortable margin for survival.

It may well be true, as today's popular spirituality insists, that our species stands on the brink of a new round of evolutionary change and that the teachings of the mystery schools have much to offer as that process shifts into gear. It is certainly true that individuals who choose to embrace the mystery teachings, and accept the disciplines of the mystery schools, can hope to meet the challenges of their own lives, as well as those of this age of the world, with a wider range of options than most of the uninitiated have at hand.

Still, neither of these hopes justify the claims too often made for them—that it is possible for individuals to transcend the limits of ordinary humanity in a single life, for example, or that we can expect a sudden evolutionary leap at some imminent moment to solve all humanity's problems. Understanding why these ideas are mistaken, and grasping the deeper hope and promise that the authentic mystery teachings offer to those willing to embrace them, will occupy the remainder of this book.

Meditation on the Seventh Law

Begin this meditation with the same posture and breathing you used to lead into the first six meditations. When you have done so, turn your attention to the theme of this meditation.

As with the earlier meditations, choose a material thing that is part of your life to be the subject of your meditation. Your task this time is to consider the whole of the evolutionary process that lies behind it—not the processes by which that individual thing came into being, but the larger history by which things of that kind have changed and developed over time. If it is something made by human hands, consider earlier forms of the same thing, or other things used in earlier ages to do whatever it is that the subject of your meditation does today. If it is something created by natural forces, consider what the same or similar forces created before the present age, starting with the relatively recent past and going back as far as your knowledge and imagination will reach. When you have a clear sense of the evolutionary process that lies behind the subject of your meditation, turn your mind's eye in the other direction, and imagine what people in the far future will use for the same purpose you use the item you are considering or what the same forces of nature will bring into being in the far future.

Work through this process with several different material things, devoting a separate session of meditation to each one. Next, explore some of the less material realities in your life in the same way. Finally, apply the same meditation to yourself; explore the evolutionary journeys that have shaped your own life—the journey that began at your birth, the journey that began with the founding of the society to which you belong, and the journey that began with the first living thing on Earth and will continue past the limits of your life toward an unguessable future.

AFFIRMATION

"Together with all other beings, I am part of the evolutionary process."

THEME FOR REFLECTION

How has your own personal evolution differentiated you from others?

THE SPIRITUAL ECOLOGY
OF MAGIC

The best way to start making sense of the limits and possibilities facing humanity in this age of the world is to explore the powers available to each of us right now. Today's popular spirituality makes some sweeping claims in this regard. A common theme among these teachings, for example, holds that the reality that each one of us experiences in our lives is entirely created by our thoughts and attitudes and, thus, can be changed completely by changing those thoughts and attitudes. As this book has already suggested, this belief rests on a partial truth—the recognition that our thoughts and attitudes do play a significant role in shaping the world that we experience—and realizing that truth can lead to other useful realizations about the ways in which we can reshape those aspects of our experience we may not find congenial.

The belief that the world is utterly obedient to the thoughts and attitudes of the individual is not simply an abstract philosophy, though, or a way of imaginatively exploring the individual's relationship with his or her environment. In much of today's popular spirituality, this belief is taken very seriously and put to work. It is tolerably easy these days to find, for instance, books and teachers claiming that limitless material wealth can be gained by changing one's thoughts and attitudes. These books and teachers usually offer instruction in simple psychological techniques—for example, repeating affirmations and visualizing one's goals as completed facts—as ways of fulfilling that promise.

It speaks well of these teachings, and of their teachers and students, that they have had the courage to put their beliefs to this sort of test. Still, the results have not exactly supported the claims. A fair number of people who practiced these methods made plenty of money during the big speculative bubbles of the recent past, but then so did plenty of other people who did not have the least interest in spirituality and did not practice any of these techniques. Furthermore, once the speculative bubbles turned to busts, a huge number of people who believed they were creating their own reality found that their reality had suddenly become unwilling to cooperate, as dreams of limitless wealth briefly buoyed by a bubble economy dissolved into a harsh new reality of economic contraction, bankruptcy, and foreclosure.

Despite this legacy of failure, the people who have pursued these teachings have grasped a principle of great importance; it is just that they have hold of it from the wrong end. A great many people feel powerless in the face of the faceless institutions and planetary crises that make up so much of our contemporary experience of the world, and no small number of them hope that spirituality might help them overcome that powerlessness and enable them to make a difference in their lives and the world. Nor are they mistaken; a great deal of the sense of helplessness that afflicts people nowadays comes from a set of habits we have learned from our culture—habits that we can unlearn, if we are willing to do so. Again, there is a kernel of truth at the center of popular notions about creating one's own reality; most of us really do have more power over our lives and circumstances than we allow ourselves to realize.

More power, however, is not the same thing as limitless power. When the false limits imposed by the cultural habits of a corrupt and crumbling society are stripped away, there remain real limits that cannot be brushed aside by a change in attitudes or by the relatively simple practices marketed in the popular literature. Even those who pursue more potent methods for bringing about change will find that there are limits to human action that no human force can overcome. Somewhere between the illusion of powerlessness and the illusion of omnipotence, the true powers of the individual human being can be discovered and used; but discovering and using these powers is a more challenging process than many people realize.

One part of the training that many mystery schools provide to their initiates focuses on the subtle powers that the human mind can exert over our experience and the ways in which these powers can be awakened and used. The traditional term that is used for these ways of action is "magic." Few words in our language or any other have been heaped with more misunderstandings. Between the mistaken attitudes that most people bring with them from popular culture on the one hand and the shrill denunciations of religious and scientific dogmatists on the other, many assumptions and misconceptions about magic have to be shed before this aspect of the mystery teachings can be understood.

What is magic? Perhaps the most useful definition describes it as the art and science of causing changes in consciousness in accordance with will. People who have no real knowledge of what magic is and what it does tend to think of magic in terms of weird rites complete with robes, wands, strange gestures, and mystic words such as "Abracadabra!" This notion is not completely wrong, since a great many symbolic objects and actions are used in magical practice, but these are the trappings of magic, not its core.

All the instruments of magic—the wands, robes, candles, billowing incense, mystic gestures, words of power, and the rest of the items in the toolkit of magical work—are simply tools by which mages (that is, people who practice magic) can cause changes in their own consciousness and in the consciousness of others. These tools are not, strictly speaking, necessary. Given enough practice and skill, a mage can get the same effects without them as with them. Still, it takes many years to achieve that level of practice and skill, and until then, the traditional instruments of magic are among the tools that most readily help mages to achieve their desired effects.

The tools of magic are useful because most of the factors that shape human awareness are not immediately accessible to the conscious mind; they operate at levels below the one where our ordinary thinking, feeling, and willing take place. The mystery schools have long known and taught that consciousness has a surface and a depth. The surface is accessible to each of us, but the depth is not. To cause lasting changes in consciousness that can have magical effects on one's own life and that of others, the depth must be reached, and to reach down past the surface, ordinary thinking and willing are not enough.

This realization marks the point where the relatively simple psychological methods used by most of today's systems of popular spirituality give way to more complex and more potent techniques. The simple methods do work, at least some of the time, and indeed, they were often given by the old mystery schools to beginners as a way to experience how the mind participates in shaping our experience of the world. It is for this reason that I have included an affirmation for each of the seven laws of spiritual ecology in the previous section of this book; by working with these affirmations, it is possible even for someone with no background in the teachings of the mystery schools to grasp something of the way the world works when seen through the eyes of these ancient traditions.

It was by way of disclosures such as these, in fact, that these techniques got into the popular spiritual literature of an earlier generation and found their way onto today's bookshelves from there. Some people, at least, have benefited significantly from these practices. Very often, though, simple methods such as affirmation, when used in an attempt to cause major changes, run afoul of unnoticed forces in the mind's depths and yield unpredictable results or none at all. To pass beyond these difficulties, the depths must be addressed in their own language, the language of dream and myth—and, thus, of symbolism.

This is what mages do when they practice magic. Every detail of a magical working—from the color of the altar cloth, the number of candles, and the kind of incense burning in the censer to the words and rhythm of the incantation—forms part of a symbolism that represents the goal the mage seeks to accomplish. That symbolism must have been studied thoroughly and made part of the self through meditation, so that the mage is able to bring surface and deep layers of consciousness into alignment and act with the power of the whole self.

Magic works, in other words, because it speaks the symbolic language of the deep self. Every action done in a magical working—be it the speaking of a word, the movement of a hand, the drawing of a breath, or the construction of an image in the mage's imagination—is a symbolic action. It means something— and something specific. In a well-designed ritual, the meanings of these symbolic acts resonate together like the notes of a musical chord, expressing a single pattern of meaning in a complete and balanced form. Seen in this light, magic is a way of unifying the self on all its levels and directing it toward a single end. This combination of unity and direction makes ritual the mage's principal tool for action on any level of experience.

Still, it is perhaps more accurate to say that this is what magic is capable of being, for it does not always achieve that goal. Reaching the point at which the possibility becomes a reality takes a great many things, but most of them are a function of one thing: practice. Those who begin the practice of magic inevitably spend most of their time fumbling with forms, memorizing words and gestures, and trying to perform rituals in the least awkward manner they can manage. Once these basic challenges are overcome, the inner aspects of the ritual have to be dealt with, over and over again, until the various levels of the rite come into harmony with one another and the first consistent effects start to show up. Even then, those effects must themselves mature and develop over time to reach their full potential.

The disciplines of magic, then, are complex and demanding, and they cannot be mastered overnight. Many of those who claim to be able to create their own reality at will, by contrast, insist that relatively simple, easily learned techniques are all that is necessary to make the world obey the individual's desires. As already mentioned, these methods do sometimes work, when the obstacles to be overcome are not very powerful and the depths of the mind happen to be receptive, and a few successes along these lines can lead the unwary to believe that they have achieved limitless power. That belief, in turn, can become a major obstacle to further progress, because it ignores the existence of the actual limits on magic's reach.

To understand those limits, it is crucial to realize that there are at least three distinct questions involved. First is the question of what aspects of our lives—what planes of existence, in the language of the mysteries—magic is able to influence. Second is the question of what the techniques that comprise the art and craft of magic actually accomplish on those planes. Third is the question of what ethical limits ought to apply to magic.

History has left the first question mired in a great deal of confusion. At one time in the Middle Ages, for example, it was considered sinful for good Christians to believe that magic could have any power at all beyond mere illusion. At a later point in the Middle Ages, it was declared equally sinful for good Christians *not* to believe that magic had power! This latter period, the age of the great witch persecutions, saw the powers of mages overstated to a preposterous degree by the propaganda machines of various churches. More recently still, of course, the pendulum has swung back the other way, and modern scientific thought considers it something close to a mortal sin to believe that magic is anything but fraud or delusion.

Amid all this obscurity, though, some things are clear. Even the most materialist thinkers should be able to recognize that magical workings can affect the whole range of phenomena that belong to the planes of mind and emotion, since the shaping of thoughts and feelings by symbols is a commonplace of human experience. Equally, those things most closely linked to the mental and emotional planes—for example, the physical body in general and the functioning of the immune system specifically—can also be influenced by magic, since symbols can affect the body by way of the mind. In terms of the Law of the Planes, the natural connection between the mental and physical planes that allows you to move your hand with a thought also allows movements of your mind to shape your health, leading to such well-known phenomena as psychosomatic illness and the placebo effect.

On the other side of the balance belong claims that magic can exert direct power over the material world on the large scale—for example, by levitating heavy objects. This sort of thing appears constantly in the fake magic of popular entertainment, and you can certainly find books and teachers insisting that someone, somewhere can make rocks fly through the air by magical means. The only problem is that no one seems to be able to do it here and now.

The Law of the Planes offers a useful explanation. The powers of magic apply directly to the planes of thought and emotion, not to the plane of physical matter, and magic can affect matter only where a point of contact exists between these planes. Living beings contain such points of contact, but as far as we know, rocks do not. Your chances of making a rock fly through the air by magic are, therefore, roughly on a par with your chances of making it weep by reading it a sad poem.

The second question mentioned above, about the technical limits of magic on the planes where it can operate, is not so much confused as ignored by most people these days. Watch what passes for magic in movies, for example, and you can expect to see the most spectacular things done by no action more demanding than waving a wand and repeating some scrap of ungrammatical Latin. In the real world, things are not so simple.

The crucial point that must be grasped here is that magic is work—very often, hard work. Even on the plane of the mind, where magic normally has its strongest and most direct effects, a single magical working may not be enough even to make a relatively simple change in consciousness. To overcome an unhelpful attitude or a self-defeating habit that is deeply entrenched in the mind can take daily practice for months at a time, and even when success is gained, patterns of thought linked to that attitude or habit may continue to surface for years thereafter. This may seem unfair, but it bears remembering that if overcoming our bad habits and useless attitudes were easy, we would all have done it long ago.

The limits of what can be accomplished by magic differ from person to person and from situation to situation. The principle here is the same as the one that governs the limits to ordinary muscular strength. Any human being past infancy, barring physical injury or illness, can lift this book; no human being, no matter how strong, can pick up a locomotive with his bare hands and carry it off on his back. In the wide spectrum between these extremes, however, there are a great many things that can be lifted by some people and not by others. Magic is subject to the same principle. Just as with muscular strength, competent training and regular practice can make a great deal of difference in one's magical abilities, but there will always be some feats that no amount of training and practice will enable a mage to accomplish.

Another limitation affects the possibilities of magic, and this limitation unfolds from the internal requirements of magic itself. A magical working must have a single purpose—not simply a vague idea of betterment, but a precise goal thought out in detail and tested against the touchstone of common sense. The Law of Limits plays a potent role here, for any purpose you choose limits your ability to seek things that are contrary to your purpose. If you desire power, for example, you can have it; if you desire love, you can have that. You cannot have both at the same time when dealing with the same person, however, because love demands vulnerability and power demands the opposite. If you try to get both at once, your workings will cancel each other, and you will get neither.

The limits of what can be accomplished by magic differ from person to person and from situation to situation.

The purpose of a magical working also needs to be measured in terms of ends rather than means, which is a philosopher's way of saying that you need to ask for what you want, not what you think will get you what you want. There's a story, well known in magical circles, of a man who tried to use creative visualization to get money. He spent hours each day imagining himself handling huge stacks of bills. Shortly thereafter, he was hired by a bank, where he spent eight hours a day, at a modest wage, counting huge stacks of other people's money.

The story is funny enough, but any number of other people have made the same mistake in less harmless ways. They have wanted security, but made money the goal of their actions; they have wanted love, but made power over someone their goal; or they have wanted peace, but made isolation their goal. In each case, they got the thing they sought and not the thing they really wanted. The way to avoid this outcome is to know yourself well enough to know what you actually want out of life, but this self-knowledge requires a degree of insight that does not always come easily.

The third question, the ethical dimension of magic, is wrapped in a fog that combines the confusion of the first question with the ignorance of the second. To begin with, the entire concept of ethics has been twisted by a set of cultural fashions, dating back a good many centuries, that define ethical behavior as avoiding as much pleasure as you can. This way of thinking has helped discredit the entire concept of ethics. Nor has the concept of ethics been helped by those popular religious movements that make a

A magical working must have a single purpose . . . a precise goal thought out in detail and tested against the touchstone of common sense.

frantic show of moral rules that nobody, not even their most fervent promoters, actually follows.

It is worth remembering that ethics once meant something very different. The word "ethics" itself comes from the Greek *ethike,* which simply means how people behave; "morals" is from a Latin word that means exactly the same thing. "Virtue," despite the prissy connotations that the word has today, comes from the same Latin root as "virility." In this older and truer sense, ethical actions consist of ways of behaving that are powerful and effective in the context of whole systems. Cowardice, for example, may save an individual's life for the moment, though it very often fails to do even this. Courage, by and large, brings better results for the individual and the community alike; that is why courage is a virtue and cowardice is a vice.

Note also that the virtues are governed, like everything else, by the Law of Balance. As the Greek philosopher Aristotle pointed out a very long time ago, courage is not the opposite of one vice, but of two; cowardice forms one end of the spectrum, foolish risk-taking the other, and courage falls halfway in between. In the same way, justice is neither selfish nor selfless, but seeks the appropriate balance between the needs and rights of self and those of others, and so on through the catalog of virtues.

More generally, all seven of the laws of spiritual ecology already discussed in this book can be understood as ethical principles as well as practical rules. Test every choice against the requirements of whole systems, the necessity of flow, the inevitability of balance, the reality of limits, the nature of cause and effect, the relationship among the planes, and the process of evolution, and you will find that any decision that makes sense when measured against these standards will be both the ethical choice and the effective choice. On the other hand, any choice that fails one or more of these tests will prove to be harmful or ineffective in the long term—no matter how expedient it seems—as well as ethically wrong.

This much can be said about any way of acting in the world, on any plane. The art of magic, though, has certain specific ethical dimensions that unfold from the way it works on the universe of our experience. Because magical action involves building up states of consciousness as a basis for its effects, it will inevitably influence the person who performs it just as much as any other person at whom it might be directed.

It is quite possible to destroy another person by magic, for example. The most common method involves building up patterns of despair, sickness, or suicide and then communicating them to the target by various technical means. The mage who does this, however, must build up these patterns in his or her own consciousness, and those patterns will manifest with as much power in the life of the mage as in that of his or her intended victim. Equally, if you set out to control the mind and emotions of another person—this is what most so-called love spells involve, of course—you will weaken your own will to the same degree that you override the will of your target.

In all this, there is no one acting as a judge; the results are simply matters of cause and effect. The same logic governs the environmental effects of pollution. If you dump poisons into the water table, you will end up drinking them, and if you pump toxins into the air, sooner or later your lungs will have to deal with them. The use of magic for corrupt or destructive ends is simply a different form of pollution; it differs from the more obviously toxic pollutants of our time solely in that it exists and acts on more subtle planes of being. If you practice magic—including such simple practices as affirmations and creative visualization—it can be useful to think of your magical work as being a smokestack that will put whatever you do into the magical atmosphere around you. Is what comes out of your smokestack something you are willing to breathe?

There is another dimension to the ethics of magical practice, one that bears directly on the core theme of this book. Magic of one kind or another has been practiced in every culture and every age, by people of every walk in life, from hunters and gatherers to housewives and farmers to priestesses and popes. Magic has also been practiced by initiates of the mysteries. Many mystery schools, however, have traditionally required their students to accept sharp limits on their magical activities during their training; some forbid students from using magic for practical purposes altogether, and the vast majority of schools have stressed that training in the mystery teachings is not a means to material wealth and power over people or the environment.

The logic behind these restrictions is as straightforward as the reasoning that governs the rules of an athlete's training. The purpose of initiation into the mystery teaching is to open up a wide range of higher potentials that lie sleeping within each human being. Central to that journey of awakening is a process of sorting out what is within the self and what is outside it. Some aspects of that process have already been sketched out in these pages, such as the recognition that many people who think they want material wealth actually desire security and self-respect, and they mistakenly think that they can get these emotional states only by accumulating money or consumer goods.

This is where magical practices can either further or forestall the work of initiation. A person who thinks he needs money to feel self-respect will normally have to work hard to see past the illusion, and he will have to recognize that self-respect depends on the state of his consciousness, not the state of his bank balance. Magical practices can be used to help that recognition on its way. On the other hand, if he directs his magical efforts instead toward getting money in order to feel self-respect, he simply feeds the illusion, wasting time and effort that could have been directed to more useful goals.

For those who are on the path of the mysteries, and also for those who seek to enter that path, the magical workings that matter are those meant to energize and illuminate the mind and consciousness of the mage. Magical workings directed toward changing the world in the service of individual desires are very often counterproductive. To borrow a phrase from today's popular spirituality, students of the mysteries set aside the attempt to create their own reality, and instead they get to work on creating themselves. It is to this work—in the language of the mysteries, the work of initiation—that we now must turn.

THE SPIRITUAL ECOLOGY
OF INITIATION

Of all the differences that separate the teachings of the mysteries from the ideas popular in today's mass-marketed spirituality, the gap between the ways these two movements think about spiritual development is probably the largest. Both recognize, to be sure, that each human individual has potentials that are rarely awakened in the ordinary course of child raising and education, but that can be developed by a special course of training in adulthood. Popular spirituality, by contrast, commonly insists that those potentials can be awakened much more quickly and easily than the tradition teachings of the mystery schools would suggest.

Sometimes this difference is taken to absurd lengths. It is not too hard to find, for example, teachers who claim to be able to turn ordinary middle-class Americans into superhuman ascended masters in a short, though usually unspecified, amount of time, by means of some exotic process that involves very little in the way of spiritual practice—though it's only fair to note that this process does commonly require a great deal of money to change hands. Even when popular spirituality does not stoop to this level, the powers it claims to grant tend to be much more spectacular than those that authentic mystery schools claim to awaken in initiates, while the time and effort supposedly required to achieve these powers are much less than what is required by the mystery schools.

This redefinition of the way of initiation has a great deal in common with the grade inflation that afflicts so many other forms of education in our society, and it arises from the same causes. Most people, given a choice between a long and difficult course of study and a shorter and easier one that claims to give the same benefits will choose the latter; it takes an unusual degree of insight to recognize that what you get out of an education in any field is exactly measured by what you put into it. Equally, if one course of study claims to offer much greater benefits than another, it can be difficult to step back and ask hard questions about the grander claims. Both these factors have encouraged teachers of popular spirituality to enlarge their claims and simplify their courses of instruction, and those who do have been rewarded with more students and larger incomes.

The proof of the pudding is in the eating, however, and the rate at which modern popular spirituality is successfully turning out immortal ascended masters is not much greater than the rate at which it is successfully enabling people to attract limitless prosperity in the midst of a sharp economic downturn. Meanwhile, the mystery schools have a long track record of producing individuals who may lack the gaudy powers promised by popular literature, but display a range of remarkable abilities on a level noticeably above the human average. How they accomplish these things and what the process of initiation reveals about the nature of humanity and the universe are the central themes in this chapter.

Over the last three centuries or so, there has probably been more nonsense written about initiation than about any other single subject, with the possible exception of the lost continent of Atlantis. The reality, as suggested in the first chapter of this book, is far simpler and, at the same time, far more interesting. Initiation consists of two phases: the first centers on ritual, the second on study and practice.

Rituals are patterns of symbolic action used to bring about specific changes in consciousness.

Rituals, as we have already seen, are patterns of symbolic action used to bring about specific changes in consciousness. In initiation rituals, the focus is on the consciousness of the candidate—the person who is being initiated—and the changes are those that prepare the candidate for a new phase of spiritual development. Because it takes a series of changes, carefully arranged like the rungs of a ladder, to move from the states of consciousness common in everyday life to those that open up the hidden powers of the individual, most mystery schools use a series of initiation rituals, which are separated by periods of study and practice. These rituals may be performed by a trained group of initiates for the benefit of each new member, or they may be assigned to the individual member to practice privately, usually many times, until the effect of the ritual is solidly established. Both methods work, and each has its strengths and drawbacks.

After each ceremony is performed, the new initiate faces the task of building on the results. This is done using the standard tools of the mystery teachings; such tools may include meditating on the symbols that play important roles in the initiation, performing rituals that reinforce the changes in consciousness set in motion by the ceremony, and studying books of spiritual philosophy that provide the context and background of that stage in the initiatory path. These practices allow the initiate to make the most of the initiation ceremony and the teachings presented along with it, and they prepare him or her for the next initiation ceremony in the sequence used by that mystery school.

If this seems like a slow and laborious process for bringing about spiritual evolution, it is. The reason it is used by the mystery schools is simply that it works, while other methods that claim to be able to rush through the doors of higher consciousness generally either do not work or have serious side effects that outweigh whatever benefits they yield. Like all other things, the mystery schools have evolved over time, adapting to changes in society and the individual and, in the process, learning which methods get the best results with the fewest problems.

The secrets of awakening the soul's hidden potentials are not actually secret at all, and they can be found in our own spirits, minds, and bodies.

The idea that spiritual evolution can and should be pursued by the sort of patient, steady effort just outlined, however, flies in the face of one of the most common notions in today's world: the idea that the inner potentials of the self can only be awakened by those who possess some sort of secret knowledge or use some secret procedure. The logic seems to be that since the way to rouse those potentials is not obvious to most of us, it must be hidden away in Himalayan temples, Egyptian pyramids, or some equally romantic setting.

This way of thinking provides plenty of opportunity for marketers, who can claim access to those hidden secrets, but it decisively misstates the nature of spiritual development. The secrets of awakening the soul's hidden potentials are not actually secret at all, and they can be found in a place that is a good deal closer to most of us than Egypt's mystery temples, though every bit as mysterious: our own spirits, minds, and bodies.

The real secrets of initiation, in fact, are the same factors that bring success in any other human activity—a point that has led more than one initiate to suggest that all human activities can best be understood as training for the mysteries. Because these factors are so often wrapped in layers of mystification when they are applied to spirituality, it can be helpful to look at the same factors at work in a different setting.

It is useful, for example, to compare the work of initiation to the challenges facing someone who wants to become a rock musician. The desire means nothing until it is expressed in action, and even then, the action must be guided by a clear sense of the work that has to be done. It is not enough to call oneself a musician, or dress like a rock star, or read books about music, or stand on a stage somewhere and fumble with a guitar while standing in the spotlight. Too much attention to these outer forms can get in the way of the work that is needed to become a rock musician. Instead, our would-be rock star needs to get competent instruction in how rock music is played. He or she also must obtain an appropriate instrument and practice on it, not just occasionally but as often as possible, until the skills and reflexes needed to play the instrument become second nature and the personal element begins to reveal itself through the mechanics of playing.

The process of becoming an initiate of a mystery school follows precisely this same course. The would-be initiate has one advantage over the would-be musician, however, because each of us is born with the only instrument we need to play the "music" of the mysteries: the human body, mind, and spirit. Still, initiates-to-be must seek out competent instruction in the mystery teachings, and they must approach the practices of a mystery school with the same persistent effort and regular hard work that a musician has to devote to music.

The regular practice of meditation, ritual, and study is, therefore, fundamental to the training of the initiate in any mystery school. Every serious student of the mysteries devotes time to these practices on a daily basis, establishing a routine that soon takes on a life of its own. This may seem counterintuitive, since part of the attraction of the mysteries is that they seem to offer an escape from the routine, the ordinary, and the predictable. Still, human beings are creatures of habit, and the mystery schools learned a very long time ago to put this fact to good use. Difficulties that cannot be overcome by willpower alone can quite often be worn down, step by step, through the deliberate cultivation of helpful habits.

The regular practice of meditation, ritual, and study is fundamental to the training of the initiate in any mystery school.

It is for this reason that most mystery schools teach their students to set aside a short period of time, perhaps half an hour, every single day for their spiritual practices. Most mystery schools also encourage their students to do their practices in the same place whenever possible and to do them in the same order. Establishing a regular routine in this way has a valuable payoff, for the student's practice time becomes an ordinary feature of his or her day; as practice time approaches, the mind automatically turns to the work at hand, and skipping a practice produces the same kind of internal upset as does skipping a meal.

Attempting to rush the gates of higher consciousness is not a good idea. Inner development is a natural process that takes its own proper time.

Another advantage of this approach to training is that nearly anyone can put half an hour a day into spiritual practices, if only by setting the alarm thirty minutes early so that the work can be done before the rest of the family wakes up. When free time becomes available, more can be put into the work of the mysteries. More is not always better; especially in the early stages of the work, the mind and body both need plenty of time to recover from the unfamiliar efforts and transformative effects of spiritual practices. This is another reason why attempting to rush the gates of higher consciousness is not a good idea. Inner development is a natural process that takes its own proper time, and attempting to force it along by exotic means is no more helpful than trying to speed up the growth of flowers by pulling them out of the ground.

The consequences of this training can be far more dramatic than a casual glance might suggest. For many centuries now, the mystery teachings have pointed out that most people tend to go through life as though they were sleepwalking. The force of habit, though it can be a potent help to the initiate's work, has a less pleasant side; it is a tyrant that rules many lives. Patterns of behavior set in motion at one point in time often continue long after they have stopped being useful or even harmless. This is equally true of the fixed habits of thinking that are usually called "opinions" or "beliefs." Tangled in unhelpful habits and unexamined beliefs like a fly in a spider's web, people too often stumble from one crisis to another, blaming the universe around them for disasters that their own actions have created for themselves.

The initiates of the mysteries, by contrast, practice the art of living deliberately, and they learn how to choose their actions and reactions, rather than to have them chosen by the power of habit or the social pressures exerted by other people. This doesn't mean that all habits of thought and action are to be abolished or that all social customs and expectations are to be ignored. On the contrary, it means that the habit-making powers of the mind and the pressures of society are to be explored, made subject to the control of the conscious self, and used when and only when they are appropriate tools for the work at hand. The initiate of the mysteries is the master of habit and social existence, rather than their slave.

The initiates of the mysteries practice the art of living deliberately.

How is this done? The critical factor here, as elsewhere in the work of the mysteries, is self-knowledge. The same clarity of purpose, strength of will, and capacity to learn that play central roles in mystery training can also be turned, like a flashlight, onto the events and actions of daily life. In practice, once these powers begin to emerge in the individual, they spread out by themselves from the realm of spiritual practice and begin to reshape the initiate's life. As this process ripens, choices that were once made automatically, in response to habit or social pressure, start being made in more productive ways, and difficulties produced by unproductive opinions and habits dissolve as those factors go away. The results quite commonly transform the daily life of the student of the mysteries beyond all recognition.

Some students of the mysteries become overenthusiastic when they realize just how sweeping the possibilities that become accessible through self-knowledge are, and they respond to the first stirrings of the process by trying to make wholesale changes in their lives all at once. Almost always, these attempts fail. This is partly because trying to act before awareness has matured adequately is a little like trying to drive a car while blindfolded, and doing so can produce roughly the same amount of wreckage.

Partly, though, these failures come about because strong habits are often responses to emotional turmoil, and changing the habit generally involves facing the emotions involved. For this reason, it's normally wisest for students to concentrate on their mystery schools' studies and practices and to allow the wider awareness and its consequences to take shape in their own time. The latter are rarely delayed for long.

One of the most striking and least understood dimensions of the mystery schools has a crucial place in the awakening of consciousness. The traditional secrecy of the mysteries was a hard necessity in those times, and they have been many, when those who wanted to follow the path of the mystery schools risked their social standing, their careers, or their lives. Those days have not passed away entirely even now. Nowadays, a variety of subcultures openly practice alternative spiritualities in the more tolerant corners of the world, but there are plenty of places even in the world's democratic nations where being openly known as a student of magic and the mysteries can put one's job in jeopardy or run the risk of harassment by local religious zealots.

At the same time, secrecy is much more than a simple means of protection. It also forms a core element in the training of the mystery schools. The discipline of secrecy teaches the initiate of the mysteries to think before speaking and to judge words and actions alike against the yardstick of a broader purpose. It is thus a powerful awareness exercise, and it also fosters the skill of living deliberately rather than habitually.

The discipline of secrecy teaches the initiate of the mysteries to think before speaking and to judge words and actions alike against the yardstick of a broader purpose.

More subtly, it brings about a change in the way the initiate relates to the social environment. That environment may be understood as a web of communication that is made up of a whole range of verbal and nonverbal actions and subject to each of the seven laws discussed in this book. Most of the flow of information that binds that whole system together goes on unnoticed, for it is a fabric woven of small talk; subtle cues; details of clothing, gesture, and voice; and the like. The very invisibility of these patterns of information is the source of their power; like the symbolic language of magic, they speak to the deep places of the mind, not the conscious surface.

Many mystery schools teach their students exercises that directly or indirectly bring this web of flowing information into conscious awareness, and they teach students ways of shaping the web deliberately. It can be instructive to discover just how unthinkingly most people react to such simple cues as the style of clothing one wears, and it can be even more instructive to discover that these same reactions influence one's own feelings and thoughts. A more powerful source of transformation, though, is the experience of breaking free of the web of communication for a time and moving outside the ordinary patterns of wordless social communication to explore the undefined space beyond it.

The web can be broken in many ways. Some of these ways have severe drawbacks; a large part of the fear so many people have of the mentally ill, for example, comes from the fact that madness pushes its victims out of the web of communication in ways the mentally ill themselves cannot control and those around them do not understand. Similarly, a good deal of the psychological damage suffered by children in abusive families comes from the culture of silence, almost universal in such families, that demands that certain subjects and incidents, however traumatic, must never be made part of the web. A good fraction of human misery has its roots in such habits.

There is a far more limited and deliberate way to break the web of communication, however, and the mystery schools have made use of it since time immemorial. A secret—any secret, about anything—stands outside the social environment by the simple fact of its secrecy. Since the fact that a secret exists at all is itself part of the secrecy of the mysteries, only the person who keeps the secret knows that a break in the web exists.

The initiate of the mysteries, as a possessor of secret knowledge, thus stands on the border of the social environment, half in the web of communication, half out. He or she can take part in the web freely, but must learn to do so in full consciousness, for a lapse into unthinking behavior risks disclosing the secret. At the same time, the secrecy of the initiate is voluntary, not forced, and covers only a small part of the initiate's life. In traditional mystery schools, the secrets students are asked to keep private often involve nothing more than a few passwords, a few symbolic gestures, and the details of a small number of rituals, yet keeping these secrets can be enough to reshape their approach to the whole social environment around them.

The practice of secrecy thus turns the very fabric of the social environment into a tool for learning and a realm of awareness and conscious action. This practice is the basis of much of the traditional image of the wizard, and many of the powers wielded by initiates have their source in their mastery of secrecy and subtle communication, rather than in any more obviously magical work. Once the countless small messages passing through the social web are made conscious, they can be read, shaped, and directed by the initiate in a purposeful way, with remarkable effects on the ecological system we call human society.

The practice of secrecy turns the very fabric of the social environment into a tool for learning and a realm of awareness and conscious action.

None of this, of course, means that the initiate of the mysteries today has to maintain the total secrecy that was needed in the days of the Inquisition. In today's less dangerous social climate, there are times and places for speech as well as for silence. Many mystery schools nowadays, accordingly, maintain a modest public presence, and members of a school are rarely forbidden to mention their involvement to others—though there will always be at least a few details of philosophy or practice that are to be left unspoken. Those who have sought the mysteries in good faith, and for the right reasons, have always been able to find them. At present, though, the relative safety given to alternative spiritual traditions in many countries makes finding them a little easier than it has been at other times.

Those who have sought the mysteries in good faith, and for the right reasons, have always been able to find them.

Those readers who are familiar with the grandiose promises so common in popular spirituality will likely be wondering, by this point, what the mystery schools can teach that can compare with the fast track to immortality and limitless power so many well-known teachers claim to be able to offer. This is a reasonable question. Do the authentic mysteries leave their initiates in the ordinary human rut, with nothing more than a bit of wisdom and a few curious mind tricks to show for their troubles, or do they offer a route to some superhuman condition?

An answer to this question is more complex than it seems, for at least three reasons. First, the limits of human potential are no more certain as the limits of magic. The potentials within us are no more limitless than anything else in the world of manifestation. Between the things any of us can accomplish and the things none of us will ever accomplish, though, lies a wide, gray area, and some of the things that fall in that area stray into realms of possibility that many people would consider superhuman. Furthermore, choice is a potent factor here, as elsewhere; initiates can always choose to use the Law of Limits to their advantage and forgo one set of possibilities in order to accomplish more in a different direction. More generally, too, there is often a great deal of room between the limits that most of us accept on our potential and the limits that are actually imposed by our nature and the nature of things.

The second factor making this a complex issue is the simple fact that every human being has different possibilities and different limitations. Each of us, according to the teachings of the mysteries, has within us a capacity for magnificence in some aspect of human life. That capacity may or may not bring fame and fortune. For every Mozart or Mother Teresa whose gifts are celebrated around the world, there are thousands whose magnificence in some modest profession or some unpaid activity will only be noticed by a few people, or by no one. There are many more in turn who let themselves be distracted from their capacity for magnificence by social pressures or misguided notions of what is and is not important, and they never manifest their inner gifts at all.

The third factor, however, is the most important. The Law of Evolution, as we have seen, does not support the notions so often piled atop evolution in today's popular spirituality or, for that matter, in popular culture generally. Evolution does not proceed by leaps and bounds, but by slow steps, according to a threefold rhythm of challenge, response, and reintegration. Each of these steps takes time—time to face the challenge, time to develop the response, and time to reintegrate the resulting abilities into the whole system of the species, the ecosystem, and the individual. Those lungfish who first ventured onto land in the Devonian period did not leap out of the water, sprout scales or fur all at once, and go capering across the landscape; it took countless small changes, each one a triumph of adaptation to the challenges of an unfamiliar world, to transform lungfish into amphibians, amphibians into reptiles, reptiles into mammals, and eventually mammals into us.

Spiritual evolution works in the same step-by-step manner. According to the mystery teachings, each life presents each soul with a challenge to which it must find an adequate response, and the soul then must integrate that response into the whole system of the self. Repeated many times over, this rhythm of awakening can transform the individual in almost unimaginable ways, opening the way to realms of consciousness and power that approach the upper limits of the human and, ultimately, surpass them.

The key word here, however, is "ultimately." Human existence is not a barrier to leap past, much less a prison to escape; every mode and expression of human life in the world is a challenge that calls for its own response or, to shift metaphors slightly, a lesson with much to teach and many possibilities to be explored. To attempt to rush through the human to reach the superhuman is to miss all the benefits to be gained from being human in the first place. It is impossible to go beyond something, in other words, without going through it. Those few among us who are actually ready to go beyond human existence show, by every one of their actions and attitudes, that they have learned all the lessons that human existence has to teach and are calmly preparing to move on.

To attempt to rush through the human to reach the superhuman is to miss all the benefits gained from being human in the first place.

The work the mystery schools set before their students, in accordance with the evolutionary sequence of challenge, response, and reintegration, thus comprises three stages. The first is a stage of preparation, in which the novice pursues the introductory course of study and practice provided by his or her mystery school, learning the basic lessons of self-knowledge and establishing the habits of study, meditation, and ritual that will set his or her feet on the path of initiation. The second is a stage of transformation, in which the student proceeds along the path step by step, learning the lessons of the mysteries and applying them ever more exactly to his or her own inner potentials and personal needs. The third is a stage of expression, in which the initiate has embodied the mystery teachings and can then apply the gifts of those teachings, along with his or her own awakened talent for magnificence, to the challenges of living in the world.

Initiation, in other words, is a natural process that follows the laws of spiritual ecology in all its details. It closely parallels the life cycle of individuals, the process of succession in ecosystems, and the process of evolution in species. In all of these cases, there is the preparation that sets the stage for the challenge; the transformation in response, which moves step by step toward its fulfillment; and the achievement of a new relationship, which reintegrates the individual into the whole system in which the process takes place. As it proceeds stage by stage, initiation tends to accumulate adaptations in the same way as other forms of evolution, but here the adaptations that accumulate are those of compassion, wisdom, and power.

Many mystery schools follow this threefold rhythm by dividing the course of initiation into three degrees or grades, each with its own initiation ritual and course of studies. Other schools have found that a more extensive set of initiations communicates their teachings and practices more effectively. Among Western mystery schools, the most common pattern involves three degrees, but it is not at all unusual to find systems of seven or ten degrees of initiation, and these are just as valid as the threefold system.

The habit of self-importance being as common among human beings as it is, the higher degrees of all these systems sometimes end up surrounded by a haze of inflated claims and overblown expectations. These are no more accurate, or helpful, than the fantasies that turn the quiet and practical work of the mystery schools into a gaudy panorama of imaginary temples and superhuman powers. Those who have reached high degrees of initiation in the mystery schools are not superhuman. If they have truly learned the lessons they claim to teach, they are generally wise, patient, capable, and strong; they have skills and capabilities many other people do not have, and they can accomplish unexpected things when the situation requires it.

Still, their work as initiates consists of teaching the mysteries and working quietly for good in their communities and societies, not of entertaining the curious and skeptical with a display of tricks or, for that matter, cashing in on the hopes and fears of the gullible. They do not promise an escape from the basic conditions of human existence, because those conditions—even the ones we dislike the most—are the limits that allow us to enter into manifestation, to manifest beauty and power, and to accomplish the work we are here to do.

It is here that the teachings of the mysteries and those of today's popular spirituality come most dramatically into conflict. A great many people nowadays, following the promptings of most of a century of popular spiritual literature, insist that the world stands on the cusp of a vast transformation that will change the basic conditions of human existence forever. The exact nature of the expected change varies drastically from book to book, teacher to teacher, and often from one believer to another. There are some who expect the world as we know it to be destroyed in one horrific process of mass extermination; others look forward to a sudden mass leap into enlightenment that will transform our troubled world into Utopia; and still others—and there seem to be a great many of them—somehow manage to combine both these images into a single vision of the future. What do the principles of spiritual ecology and the teachings of the mysteries offer in response to the passionate hopes and fears that cluster around such ideas at this turning of history's wheel? The final chapter of this book will explore that theme.

THE SPIRITUAL ECOLOGY
OF HISTORY

It is one of the most interesting features of today's popular spirituality that so much of it should be fixated on the imminent end of the world we know. Our time is not quite unique in this obsession, but it definitely falls into a small minority among the ages of the world. There have been other times, to be sure, when most people have waited breathlessly for the world as they knew it to come to a sudden stop. Far more often, however, people have assumed that the world would continue in its familiar path, and they have been right. In those calmer times, the task of spirituality—and thus of the mystery schools—has been understood as the inner transformation of the individual and the gradual improvement of society.

In every generation, though, there have been at least a few people who have insisted that history as we know it was about to end and be replaced by something more closely attuned to humanity's fears, its desires, or both at once. Great civilizations at the height of their wealth and power have seen prophets stalking the streets with dire warnings of the wrath to come; tribal peoples overwhelmed by invaders from overseas have turned hopeful ears to visionaries who promised salvation at the hands of supernatural forces. Great poets and artists have embellished these visions, mass movements have followed them, and they have sometimes turned out to have a tremendous impact on the history of their time. Yet it needs to be remembered that all these prophecies had something important in common. Every one of them, without a single exception, has turned out to be wrong.

This last statement needs to be understood in perspective. History is full of happenings that could, with a big enough helping of poetic license, be described as apocalyptic. For the people who were living in the city of St. Pierre on the island of Martinique when a volcanic eruption destroyed it in 1902, leaving only two survivors out of a population of thirty thousand, it is not unreasonable to say that the world ended. It is just as reasonable to describe the fate of the inhabitants of Hiroshima on August 6, 1945, in the same terms. History is full of such events, and the destruction of a city, a country, or an entire civilization is hard to describe without falling back on images not far from those of the apocalyptic prophecies just discussed.

Still, there is an important difference between even the most terrible event in actual history and the claims of total planetary transformation that fill today's popular spirituality. While the eruption of Mount Pelée and the mushroom cloud over Hiroshima meant the end of the world for many individuals, for the survivors and those living elsewhere, life went on. For everyone else who was still on Earth, in other words, there were still dishes to wash, livings to earn, and the ordinary limits of the human condition to cope with.

Furthermore, in the aftermath of these catastrophes, as in every other age of history, those who wanted a better world or who sought the inner transformations that the mystery teachings can provide had to work for either one of these goals one slow step at a time. This is the unwelcome reality that prophecies of apocalypse always claim to escape in theory, but never manage to evade in practice.

The attempt to evade unwelcome realities, in fact, underlies most of the popular habits of thinking this book has discussed, and it forms one of the great contrasts between these ideas and the authentic teachings of the mysteries. Most of today's popular culture, in and out of the spiritual field, insists that the troubles we face in our lives can be banished without the hard work and self-discipline that the mystery teachings demand. Today's fashionable philosophies insist that the leap to a higher and happier mode of being can take place all at once, by some means other than personal effort and insight. Fairly often the method proposed for making this leap bases itself, in one way or another, on the claim that the unwelcome aspects of human existence will go away if we all simply pretend that they do not actually exist.

Today's fashionable philosophies insist that the leap to a higher and happier mode of being can take place all at once, by some means other than personal effort and insight.

This approach to the human condition runs up against a good many difficulties, but the most important of them has been addressed more than once already: when put to the test, these ideas fail to perform as promised. This awkward fact puts believers in a difficult bind. They could, of course, accept that the ideas they have embraced with such enthusiasm do not happen to work, but this is a bitter pill for most people to swallow. It is made more bitter still when, as so often happens these days, the people who have embraced these beliefs have gone around proclaiming them to other people and, thus, face a great deal of social embarrassment if they have to admit their mistake.

Fairly often, faced with these unpalatable choices, believers respond to their crisis of faith by finding some reason to keep on believing, even in defiance of the facts. Having failed to attract prosperity into their lives, for example, they may decide that this happened because they didn't trust the teachings enough, or they come up with some other reason to put the blame on themselves rather than on the teachings. Taking the blame themselves very often leads them to seek ways to transform themselves into the kind of people who can make the teachings work. That search occasionally brings them to the doors of the mystery schools, but more often it sends them in search of some less strenuous option. Today's popular spirituality meets them more than halfway by offering a range of colorful empowerments and ceremonies that promise to give them the power to make the world conform to their beliefs.

When these latter attempts fail in their turn, as they inevitably do, and the unwelcome realities of human existence remain fixed firmly in place, the next step is the one that gives this chapter its theme. When the world refuses to conform to the desires of the believer, one ancient and widespread gambit is to insist that the world that makes this refusal will shortly be replaced, with or without apocalyptic fireworks, by a world in which the cherished belief system will work as advertised and in which everyone who currently disagrees with that system will be forced to admit that they were wrong. Another popular gambit is to look for scapegoats to blame for the world's failure to follow the belief system, on the assumption that everything would work out as promised if only some nefarious group or other was not busily interfering with the way things ought to be. Fairly often these two themes fuse, creating lavish and exciting narratives in which a hopelessly wicked world ruled by evil forces—and, thus, annoyingly unwilling to behave the way some belief system insists it ought to do—is about to be destroyed utterly by an apocalyptic event and replaced by a new and sparkling world from which everything that contradicts the cherished belief system will be excluded forever.

This trajectory can be traced all through today's culture. The evangelical Christian revival that began in America in the 1970s, for example, has followed it every step of the way from its bright beginnings to its bitter end. The enthusiastic days of the "Jesus freaks" and the Good News Bible, when countless young people turned to Christianity as a way out of the unsolved conundrums of the sixties, faltered as it turned out that believing in Christ did not manage to provide easy answers to life's hard questions. There followed an era of increasingly strident claims that believers could suddenly and easily be "born again" into a new life of perfect happiness in Christ, if only they believed passionately enough. Finally, of course, the movement has completed its trajectory, and many people who once dreamed of a world awash in Christ's infinite love now cling to the hope that their savior will shortly rapture them away from a world of monstrous evil, which they believe is ruled by wicked and powerful elites who are supported and directed by Satan himself.

In the same way, many of the same New Age circles that twenty years ago enthusiastically discussed the individual's power to create his or her own reality, and ten years ago were full of talk about becoming an ascended master in this lifetime, have now turned much of their attention to garish prophecies of cataclysmic earth changes and a dizzying assortment of conspiracy theories that blame the failure of New Age ideas to live up to their promises on anything and everything but the belief systems themselves. This transformation seems to have taken place without anyone asking why people who create their own reality would choose to create one that is ruled by forces of absolute evil and is poised on the brink of mass death. Still, this shift should come as no surprise. The same road has been walked many times before, by many different peoples and cultural movements, and the only startling thing about it is the fact that those who travel on that road always insist that they are doing something new.

The teachings of the mystery schools understand history in a very different way. That way can best be understood through the same Law of Evolution that governs the spiritual ripening of the individual soul. Each human society arises out of the chaos left behind by some previous society, and it takes shape in response to whatever challenge the older society could not meet. As it emerges, the newborn society struggles to adapt to the challenge that brought it forth, and it finally achieves a stable adaptation that will allow it to reintegrate into the whole systems that surround it. It endures for as long as environmental conditions and its own limitations allow before finally disintegrating in the face of some new challenge that brings a new society into existence in turn.

These cycles of challenge and adaptation can sometimes add to the sum total of human knowledge and insight, and in this sense, at least, human societies can be said to progress. Still, it is unfortunately common for the people of one society to ignore the hard-won wisdom of older societies and to suffer as a result. Our own age is a case in point. Modern industrial society has parlayed its mastery of a handful of technical tricks into the most complex technology in recorded history, and it has learned many things about nature that other human societies never knew. It has achieved these things, however, at the cost of ignoring many lessons about humanity and the world that other ages knew well.

Our future will likely follow the common trajectory of civilizations in decline.

Our blindness to our own dependence on nature's cycles and resources is one result of this ignorance, and that blindness promises to make the lifespan of industrial society far shorter than that of many other human societies of the past. This does not mean that we are facing the sort of apocalyptic fantasy so popular these days. What it means instead is that our future will likely follow the common trajectory of civilizations in decline, though that trajectory will be made somewhat rougher by the side effects of the very technologies that helped us rise. As the ancient philosopher Heraclitus taught, "The way up is the way down"; it took some three centuries for modern industrial society to rise out of its preindustrial origins, and it could well take another three centuries for it to decline to a postindustrial world, where many of the technologies that now fill our lives will have become the stuff of legend.

The teachings of many of the mystery schools hold, curiously enough, that something very like this happened to human beings long ago. As mentioned earlier, an extraordinary amount of nonsense has been written down through the years about the lost continent of Atlantis. Sweep away the wilder speculations and the more obviously symbolic tales, though, and the old story bears a crucial lesson for us here and now.

The Atlantis legend claims that humanity, at one point in the distant past, created a society as complex, as powerful, and as arrogant as our current civilization. It also claims that this society attempted, as we are attempting, to make the world obey its desires without paying any attention to its own dependence on the cycles of nature. According to the legend, the people of Atlantis ignored the laws of spiritual ecology, and by the time those laws finished with them, nothing remained of Atlantis but gray waves rolling across empty ocean.

The people of Atlantis ignored the laws of spiritual ecology, and by the time those laws finished with them, nothing remained of Atlantis but gray waves rolling across empty ocean.

This legend has given rise to a great deal of speculation about whether an advanced human civilization flourished toward the end of the last ice age and drowned beneath rising waters as the great glaciers melted. A certain amount of evidence, none of it conclusive, supports that possibility. Intriguing as such speculations may be, they are not really the point of the legend. Its value, rather, is as a cautionary tale—a reminder that, among other things, people who think that they are pursuing their own best interests can be doing exactly the opposite without realizing it.

In a way, the rise and fall of the recent housing bubble was an Atlantis in miniature. Millions of people in America and elsewhere convinced themselves that they were destined to get rich without effort, and instead their actions saw to it that most of them became much poorer than they otherwise would have been. It is a fine bit of irony that those people who have ended up owing more money on their homes than the homes are worth are called, in today's real estate jargon, "underwater."

The mystery teachings suggest that the sinking of Atlantis was a gradual process, involving several epochs of flooding interrupted by periods of stability.

However difficult it was for those who lost their homes and their savings, though, the end of the housing bubble was not the end of the world. Neither, according to the legend, was the end of Atlantis. The mystery teachings on the subject suggest that the sinking of Atlantis was a gradual process, involving several epochs of flooding interrupted by periods of stability. Each time, as the sea rose and the waves rolled in, many people died and much precious knowledge was lost. Those who survived each round of sea-level rise and had the presence of mind to realize that the situation was only going to get worse had the opportunity to take ship to distant lands, and they carried with them the seeds of civilizations not yet born. In their new homes, they faced all the same hopes and fears, limits and possibilities, that human beings face in their lives today.

The teachings of some mystery schools include prophecies that, someday in the future, a fate like that of the legendary

Atlantis will overwhelm the proud towers of our own civilization. Still, the principle just outlined applies to our future, just as it did to these legends of the distant past. Even if the oceans rise for us as they did for Atlantis and great waves someday roll over our coastal cities as, the legend has it, they once rolled over the City of the Golden Gates, the work of the mysteries will still need to be done in the same way that it is done today. Even the fall of a civilization—dramatic as it may be—does not change the nature of human consciousness or the hard work needed to develop its potentials through study and practice.

All this can be understood clearly if the laws of spiritual ecology are kept in mind. Civilizations, like other living things, are born, pass through their life cycles, and die. While they live, their actions help shape the biosphere for good or ill, and they commonly leave behind legacies that can influence other living things for a very long time after they are gone. The laws of ecology limit the harm as well as the good that civilizations can accomplish, and they guarantee that much of whatever harm those civilizations do will inevitably circle back onto themselves.

Even the fall of a civilization does not change the nature of human consciousness or the hard work needed to develop its potentials through study and practice.

Those same laws guarantee that however traumatic the fall of a civilization turns out to be, and however destructive the legacies it leaves to the future, those challenges will call forth adaptive responses and start new currents of evolutionary change. If Atlantis was a historical reality, the survivors of its fall picked up the pieces of their lives and started over in distant lands, as survivors of so many more recent catastrophes have had to do. If our own civilization destroys itself in some similar way, the survivors of that process, in turn, will face some form of the same unromantic task. Their experiences, on the other hand, will doubtless have little in common with the utopian fantasies that have come to fill so much space in today's popular spirituality.

The laws of ecology limit the harm as well as the good that civilizations can accomplish.

Those fantasies, ironically, have at least some of their roots in the prophecies and legends circulated by the mystery schools—or rather in what has been made out of those prophecies and legends after they were pulled out of their original contexts and reshaped to echo the fears and daydreams of a contemporary audience. Just as the useful insight that each of us participates in creating the reality we experience has been distorted into the insistence that we each create our own reality, the undeniable fact that each of us has the power to help make a better world by our thoughts, words, and deeds has been twisted into an insistence that a perfect world is about to be handed to us by some cosmic force or other.

Every few years, as a result, another date is proclaimed as the day of the imminent great change, and every few years, another date rolls past without ushering in anything but another round of business as usual. This is exactly as it should be, because the "business as usual" of any society, in any age, is not some sort of annoying obstacle in the way of a spiritual life. It is a necessary part of the challenge to which a spiritual life is the appropriate response. While it is partly created by the thoughts, words, and actions of the people who live in that age and participate in that society, it is also shaped by the legacies of the past and the activities of people and other living things outside the limits of that society, and it exists within limits determined by the nature of manifestation and the laws of the natural world. These latter factors cannot be waved aside. They are realities as solid as any power the human mind can wield.

Faced with the perennial mismatch between the higher ideals of humanity and the often disappointing realities of life in the world of manifestation, each of us has several choices. We can go about our lives in the usual way, settling for the world as it is and trying to get as much pleasure and profit out of it as possible before we grow old and die. We can turn to ordinary religion in the hope of consoling ourselves with the hope of a better world on the far side of death. We can take a more active role and become involved in social movements or political activism in an attempt to make the world a better place, or we can convince ourselves that changing our own thinking will have the same result on a personal or collective scale. Finally, and least productively, we can sit and wait for some apocalyptic event to make the world suddenly live up to our expectations.

All these are common ways of living in the world, and each of them has lessons to teach. They are as much a part of our collective human existence as birth or death. Those who are drawn to these paths will follow them, now as in every previous age, and they will learn those lessons in proper time. Yet there is also another way: the way of the mysteries.

This way starts by realizing that our everyday life in the world of manifestation, here and now, exactly as it is, is a lesson to be studied and understood, rather than a trap to be escaped or an illusion to be ignored. It goes on to recognize that the same laws that shape our ecological relationships with the world around us also define our existence in the subtler realms of mind and spirit and that learning to live and act in harmony with the laws of ecology—spiritual as well as material—is a core element of life's lessons. It proceeds to turn those lessons to good use in the slow process, step by step, life by life, that takes each individual soul along its own evolutionary path toward the reintegration of the individual and the universe.

It is not for everyone, this path of the mysteries. Those whose interests center on getting or keeping some particular standard of living, with its privileges and comforts, usually find it unappealing at best. Those who insist that the problems of human existence are to be solved by changing the world around us, whether by physical action or the roundabout means offered by today's popular spirituality, have typically found it incomprehensible. For those who hear the call, however, the doors of the mystery schools are open today as they were in centuries past. No matter what changes shake the world, as our civilization comes belatedly face to face with the consequences of its own mistaken decisions, the patient work of the mysteries and the gradual ripening of human potential remain what they have always been.

For those who hear the call, the doors of the mystery schools are open today as they were in centuries past.

AFTERWORD

It is not the purpose of a book of this sort to attract potential students to any one mystery school. In most countries of the modern industrial world, there are dozens or hundreds of such schools, and in the United States there are several thousand, including traditions from every corner of the world. As a student of the mysteries, I have worked with close to a dozen schools in two Western traditions, but this is hardly enough to make recommendations. It is enough, at most, to learn that the school that satisfies the inner needs of one student will as often as not be wholly inappropriate to another.

Some advice, however, can be offered to those who feel themselves called to the work of the mystery schools. It is probably necessary to mention, first of all, that many organizations of the kind I have called "mystery schools" do not use this venerable name for themselves, at least in public. Some of the organizations that do use this term for themselves are moneymaking schemes or groups purveying the sort of popular spirituality called into question in this book. A mystery school worth the name, however, can be known by certain signs, of which the following are the most important.

First, an authentic mystery school offers its teachings as inexpensively as its circumstances will permit. The realities of the world of manifestation in a materialistic age do not always permit the teachings to be given away for nothing; instructional materials cost money to print and mail, and teachers of the mysteries also have to eat. Still, any school that charges enough to place a continuing strain on even a very modest monthly income should be avoided. It is a good sign when a school's elementary teachings can be found in print or in some other form the general public can easily obtain.

Second, an authentic mystery school does not call attention to itself, beyond the modest level of publicity necessary to allow interested persons to find it: a few ads in periodicals, a website, a few pages in the back of an instructional book sold to the general public, and the like. This same modesty extends throughout the activities of a real mystery school. For example, you will never hear an authentic mystery school claim to be the only valid source of initiation or spiritual teaching, or even the best such source. Claims that the teacher of a school has unique spiritual powers or status that transcends that of other mystery teachers is a sign that the school is one to avoid. The initiates of an authentic mystery school know better than to make such claims or allow them to be made.

Third, an authentic mystery school makes only those promises it can keep, and it is able to provide reasonable proof of its claims. If you encounter a school that claims to teach its initiates how to fly through the air by magical means, for example, it is perfectly reasonable to ask for a demonstration and equally reasonable to walk away if one is not forthcoming. If a school claims to produce initiates who are wise, compassionate, and insightful, it is only fair to expect its initiates to behave that way under most circumstances and to expect them to recognize their mistakes and correct them when, being human, they occasionally fail to behave this way. To use a popular phrase, the initiates of a mystery school should always be expected to walk their talk.

Those who hope to become students of a mystery school, in turn, should expect to show the equivalent qualities themselves. They should be willing to contribute their share to meet the reasonable expenses of the school. They should be modest about their own knowledge, whatever that may happen to be, and go into the school with the assumption—a surprisingly rare one these days—that when the school's teachings disagree with their current opinions, the possibility that the school may be right should at least be worth considering. They should expect to walk their own talk, keep whatever agreements and promises they make as part of their training, and devote honest effort to the studies and practices the school offers. Mystery schools exist to teach, but they are not obligated to teach those who are not willing to learn.

When the student is ready, says an old but accurate maxim, the teacher will appear. There are plenty of things a prospective student can do to approach that state of readiness. The most important of these preparatory tasks is to learn and regularly practice some system of meditation. The meditations given earlier in this book offer a first step in that venerable art, but only a first step. There are plenty of books in print that cover the basics of meditation in a more detailed manner, and classes can also be found in many communities.

Books on the teachings of mystery schools are also worth reading, both as an introduction to the mysteries and as an important part of the early stages of following their path. Among the books I can recommend, in the two mystery traditions in which I have some experience, are the following.

William Walker Atkinson (writing as "Three Initiates"), *The Kybalion:* All but forgotten today, Atkinson was an influential teacher of the mysteries in the early twentieth century, writing under several pen names. His most important book, *The Kybalion,* has been the standard introduction to mystery teachings in American mystery schools from its publication in 1912. Its "seven Hermetic principles" should be compared to the seven laws of spiritual ecology presented in this book.

W. E. Butler, *Apprenticed to Magic* **and** *The Magician: His Training and Work:* Butler was a student and initiate of Dion Fortune (see below) and an influential teacher and founder of mystery schools. These are the best of his books, but all his writings deserve careful study.

Dion Fortune, *The Mystical Qabalah* **and** *Sane Occultism:* Violet Firth, to use her real name, was among the twentieth century's most important mystery-school teachers. She taught and initiated most of a generation of teachers of the mysteries in Britain and also wrote excellent books, of which these two are the most useful for beginning students.

Manly Palmer Hall, *Lectures on Ancient Philosophy* **and** *Self-Unfoldment by Disciplines of Realization:* The most influential mystery-school teacher in twentieth-century America, Hall is best known for his early work *The Secret Teachings of All Ages,* but the two books listed above are more valuable guides to the theory and practice of the mysteries.

William Quan Judge, *The Ocean of Theosophy:* Theosophy was founded in 1875 by a group of initiates in an effort to make the mystery teachings more available to the public of that time, and the works published under its name have been standard reading in most mystery schools in the Western traditions for many years. Judge was one of the movement's original members, and this book is the best short introduction to its teachings.

Gareth Knight, *A Practical Guide to Qabalistic Symbolism* **and** *The Secret Tradition in Arthurian Legend:* Another of Dion Fortune's students and a founder of several mystery schools active throughout the English-speaking world, Knight has written many books on the mysteries, but these two are arguably his best.

Eliphas Lévi, *Transcendental Magic:* Alphonse Louis Constant, to use his real name, played a crucial role in reviving the Western tradition of the mysteries in the middle years of the nineteenth century, when they were at a low ebb. Of his many books, this is widely considered the best.

Ross Nichols, *The Book of Druidry:* Nichols, a British poet and painter, was a major figure in the twentieth-century Druid mysteries, and this challenging but important book is a comprehensive study of Druid mystery teaching.

Lewis Spence, *The Mysteries of Britain:* A leading Theosophist in his time and the author of many excellent books, Spence was also active in the Druid mysteries and wrote this book as an introduction to them.

Rudolf Steiner, *How To Know Higher Worlds* and *A Way of Self-Knowledge:* Steiner was also active in the Theosophical movement for some time, but left to found his own mystery teaching of Anthroposophy. His writings are philosophically rich and full of valuable practical insights.

In addition, a book I coauthored with **Clare Vaughn and Earl King Jr.,** *Learning Ritual Magic,* may be useful for readers who want to know more about the Hermetic tradition of the mysteries, and my books *The Druidry Handbook* and *The Druid Magic Handbook* may be of value for those who are interested in exploring the Druid mysteries.

ABOUT THE AUTHOR

Patrick Claflin

John Michael Greer has been a student of the occult traditions for more than thirty years. The current Grand Archdruid of the Ancient Order of Druids in America (AODA), he is also a longtime Golden Dawn initiate and scholar of the Western Mystery Traditions. Greer is the author of numerous articles and eighteen books, including *The Druidry Handbook*, *The Art and Practice of Geomancy*, and *The Long Descent: A User's Guide to the End of the Industrial Age*. He is also the co-author of *Learning Ritual Magic* and *Pagan Prayer Beads*. He lives in Maryland with his wife Sara. Visit him online at: *thearchdruidreport.blogspot.com*.

OTHER BOOKS BY JOHN MICHAEL GREER

Praise for *The Long Descent*

"The Internet writings of John Michael Greer—beyond any doubt the greatest peak oil historian in the English language—have finally made their way into print. Greer fans will recognize many of the book's passages from previous essays, but will be delighted to see them fleshed out here with additional examples and analysis. *The Long Descent* is one of the most highly anticipated peak oil books of the year, and it lives up to every ounce of hype. Greer is a captivating, brilliantly inventive writer with a deep knowledge of history, an impressive amount of mechanical savvy, a flair for storytelling, and a gift for drawing art analogies. His new book presents an astonishing view of our society's past, present, and future trajectory—one that is unmatched in its breadth and depth."

—Frank Kaminski, *SeattleOil.com*

Praise for *Apocalypse Not*

"Archdruid Greer carefully describes the 'apocalypse meme' and relates its sad history since its origins in Persian Zoroastrianism 5,000 years ago. The book is an absorbing and entertaining read."

—*FATE Magazine*

"This sweeping survey of apocalyptic thought during the last three and a half millennia is written with erudition and sprinkled with humor. John Michael Greer seamlessly weaves the threads of religious/mystical and secular/revolutionary apocalyptism—from the most well-known exemplars to the delightfully obscure."

—James Wasserman, author of *The Temple of Solomon: From Ancient Israel to Secret Societies*

"*Apocalypse Not* is a riotous romp through the history of the human imagination. Greer delightfully details our inherent human need to seek a utopian world, available only to the worthy by a trial that makes it worthwhile. Reading this book will make you laugh at human folly and cry at its consequences, along with some of the most colorful figures in history."

—Jeff Hoke, author of *The Museum of Lost Wonder*

"Apocalypse NOT! It was very important that somebody took on this faux religious madness, this yearning for the death of our planet. Hooray for John Michael Greer for skewering the lot of them on his acerbic pen! From the artificial counting of the dates of the end of the world, to the last big bout of robbing people of their money, hopes, and religious beliefs, Greer tracks this male madness faithfully. This seems to be a very American obsession, the end of days, profitable to those who spread it, a doom to those who drink their Kool Aid."

—Z Budapest, author of *Celestial Wisdom* and *The Holy Book of Women's Mysteries*

"The perfect hangover cure for the day before the day after the day the world didn't end."

—Lon Milo DuQuette, author of
The Key to Solomon's Key

"If peddling flesh is the world's oldest profession, John Michael Greer makes a good case that peddling fear's not too far behind. *Apocalypse Not* lucidly spells out how social upheaval—as well as plain old boredom and frustration—have always inspired fantasies of The Great Reboot, when lions will lay down with lambs, streets will be paved with gold, and lowly stock boys will become lofty CEOs. This book explains the hows and whys of such grand fantasies throughout history and how often they seem to come to grief. Armageddon through to you?"

—Christopher Knowles, author of *Our Gods Wear Spandex* and *The Secret History of Rock 'n' Roll*

"*Apocalypse Not* breaks open the doomsday clock, revealing all its cogs and inner workings. The end isn't near: It's Greer."

—Clint Marsh, author of *The Mentalist's Handbook*

Praise for *The Druidry Handbook*

"This is an altogether admirable book that wears its author's considerable scholarship lightly. A valuable addition to any collection of books on the history of Druidry and especially the modern druidic revival, it goes far beyond this in being both a deeply spiritual and eminently practical book."

—John Matthews, author of *Secrets of the Druids*

"In this book, with great clarity, Greer performs alchemy. He discovers and articulates the gold that lies hidden within the obscure texts of the Revival Druids, and succeeds with consummate skill in offering a perspective that redeems much of modern Druidry—revealing it to be a heritage around which we can grow and build a vital and dynamic spirituality."

—Philip Carr-Gomm

Praise for *The Art and Practice of Geomancy*

"This definitive book on geomancy belongs on every magician's bookshelf. John Michael Greer presents geomancy as the Western equivalent of the Eastern I-Ching—a binary code that is deceptively simple yet capable of great profundity. Greer brings texts long ignored into the light of day, making antique concepts accessible to modern diviners and magicians."

—Mary K. Greer, author of
21 Ways to Read a Tarot Card and *Tarot for Your Self*

"John Michael Greer has taken esoteric lore that too few of us have any working knowledge of and presented it in a thoroughly practical, engaging, spiritual, and insightful manner."

—Christopher Penczak, author of
The Living Temple of Witchcraft Volumes I and II

"John's magnificent book is the greatest comment on geomancy ever written."

—Lon Milo DuQuette

Praise for *The Ecotechnic Future*

"Greer's work is nothing short of brilliant. He has the multidisciplinary smarts to deeply understand our human dilemma as we stand on the verge of the inevitable collapse of industrialism. And he wields uncommon writing skills, making his diagnosis and prescription entertaining, illuminating, and practically informative. Not to be missed."

—Richard Heinberg, Senior Fellow, Post Carbon Institute
and author of *Peak Everything*

"The enormous virtue of John Michael Greer's work is that his wisdom is never conventional, but profound and imaginative. There's no one who makes me think harder, and *The Ecotechnic* **Future** pushes Greer's vision, and our thought processes in important directions."

—Sharon Astyk, farmer, blogger, and author of
Depletion and Abundance and *A Nation of Farmers*

"In *The Ecotechnic Future*, John Michael Greer dispels our fantasies of a tidy, controlled transition from industrial society to a post-industrial milieu. The process will be ragged and rugged and will not invariably constitute an evolutionary leap for the human species. It will, however, offer myriad opportunities to create a society that bolsters complex technology which at the same time maintains a sustainable interaction with the ecosystem. Greer brilliantly inspires us to integrate the two in our thinking and to construct local communities which concretely exemplify this comprehensive vision."

—Carolyn Baker, author of *Sacred Demise:
Walking the Spiritual Path of Industrial Civilization's Collapse*

Praise for *The Druid Magic Handbook*

"Greer takes many of the ideas and techniques of Druidry and, using his encyclopedic knowledge of Western occultism, combines them to create a powerful way of making magic. With great clarity he explains the purpose and technology of this Druidic magic and succeeds in presenting a system that is thoroughly modern—even though it is rooted in the past. The world needs more magic, and more books like this which show us how we can re-enchant our lives and the land around us."

—Philip Carr-Gomm, author of *The Druid Plant Oracle*

"An intriguing and well-constructed blend of Druidry and magic. There is enough wisdom here to keep most practitioners busy for a long time. Practical, thought provoking, and timely, this is a book that should be in the collection of every practicing Druid and magician."

—John Matthews, author of
Taliesin: Bardic and Druidic Mysteries in Britain and Ireland

TO OUR READERS